The Data Book

Collection and Management of Research Data

The Data Book

Collection and Management of Research Data

Meredith Zozus
Department of Biomedical Informatics
University of Arkansas for Medical Sciences
Little Rock, Arkansas

CRC Press is an imprint of the
Taylor & Francis Group, an informa business
A CHAPMAN & HALL BOOK

CRC Press
Taylor & Francis Group
6000 Broken Sound Parkway NW, Suite 300
Boca Raton, FL 33487-2742

Printed on acid-free paper

International Standard Book Number-13: 978-1-4987-4224-5 (Hardback)

Library of Congress Cataloging-in-Publication Data

Names: Zozus, Meredith Nahm.
Title: The data book : collection and management of research data / Meredith Nahm Zozus.
Description: Boca Raton : CRC Press, 2017.
Identifiers: LCCN 2016045625 | ISBN 9781498742245
Subjects: LCSH: Research--Data processing. | Statistics. | Statistics--Data processing
Classification: LCC Q180.55.E4 Z69 2017 | DDC 001.4/22--dc23
LC record available at https://lccn.loc.gov/2016045625

Visit the Taylor & Francis Web site at
http://www.taylorandfrancis.com

and the CRC Press Web site at
http://www.crcpress.com

Printed and bound in the United States of America by
Edwards Brothers Malloy on sustainably sourced paper

Writing this book was often an exercise in generalizing from practice and connecting those concepts to tested theories from related disciplines such as cognitive science, mathematics, information science, statistics and engineering. My 20 years in managing research data were spent alongside many seasoned experts from academia and industry alike from whom I learned a great deal and to whom I owe a great debt of gratitude.

I dedicate this book to these many individuals from whom I have learned so much.

Contents

Preface

The goal of this book is to present the fundamentals of data that have, or should have, a bearing on their collection and processing. In recent years, with the National Institutes of Health and the National Science Foundation in the United Sates, and the Organisation for Economic Co-operation and Development internationally, there has been an increased emphasis on sharing data from publically funded studies and on research replication and reproducibility. What I have also noticed is a paucity of emphasis on how the data were collected and processed—yet, these are the tasks that largely determine data quality.

Most research-oriented master's and doctoral degree programs include courses on research design and analysis, yet almost none include a course or portion thereof covering the basics of data collection and management. Thus, today's investigators and research teams receive very little, if any, training in the proper collection and handling of data for research. Without some knowledge of these things, investigators are ill-prepared to demonstrate that data are capable of supporting research conclusions, to support reuse of data, and to facilitate research reproducibility and replication.

Ideally, every investigator, statistician, and research data manager would command basic knowledge enabling them to confidently match a research scenario with appropriate methods and tools for data collection, storage, and processing. For many reasons, including the ever-increasing knowledge in scientific disciplines and already tight curricula, the ideal is often out of reach. This book lays bare the fundamental concepts necessary to analyze a research scenario and understand the type and extent of data management tasks required and appropriate methods. The intended audiences include (1) graduate students and researchers who are knowledgeable in their particular field but with little practical experience in working with data and (2) members of research teams involved with the collection and management of data.

This is not a reference manual. There are no recipes that can be blindly applied to data collection and processing for research. A multimillion dollar system may be required for one research scenario, whereas a simple spreadsheet may suffice for another. Instead, this book teaches the language and strategies of data management, as well as key pitfalls. Examples of how principles may be put into operation on simplistic example research studies are provided as an illustration of how the principles work in practice.

There are many resources for curation, sharing, and stewardship of data following a research project. A reader interested in these postpublication activities should consult one of these resources. The focus of this work is on planning, executing, and evaluating data collection and processing of data for research—the activities that occur before and during the preparation for analysis and publication.

I have managed the data for research for more than 20 years in settings, including engineering, life sciences, policy research, and clinical studies. I have worked with individual investigators and small teams, and have led large institutional data management operations. I frequently observed individuals in many different roles: investigators, statisticians, project managers, administrative personnel, data managers, and others handle data collection and processing tasks in research. What I found in common was that there was, with a few exceptions, little formal training for individuals collecting and processing data for research. As I completed my own doctoral work in informatics, I began to investigate and seek the theoretical, conceptual, and methodological underpinnings relevant to the management of data in research settings. My goal was to faithfully translate these gems—often from other disciplines—to managing data for research. In many areas of practice, I found no such foundations. Where this was the case, I made my best and humble attempt to identify key concepts and generalize from practice their fundamental relationships.

I wrote this book as a single author text to maintain a consistent focus and to weave the major themes such as traceability, assurance, and control throughout the text, as these are central to data management. It was in deference to these that the increased richness of a multiple author work was sacrificed. I hope that what this book lacks in perspectives from the many experts in the field it makes up for as a streamlined articulation of key principles and frameworks.

The progress of the human race relies on research results. The data are the foundation on which those results stand. It is important that handling the data is approached with the same sense of rigor, value, and discipline as the research itself.

Introduction

As our command of technology and computation increases, so does our ability to collect data and lots of them. Today, data are available directly from measuring devices, from everyday objects such as cars and coffee pots and everyday transactions such as shopping and banking. Our ability to probe subatomic, atomic, molecular, and cellular phenomena continues to increase as does our ability to observe phenomena outside our solar system and indeed the galaxy. To science, new measures mean new insights.

To investigators and research teams, these riches mean more data, from more sources, and more quickly. In this way, these riches mean that most areas of scientific inquiry are becoming more data intensive. All of these data must be defined, organized, processed, stored, and computed over. Thus, most investigators' qualitative and quantitative probing phenomena on the smallest to the largest scales are increasingly reliant on data and can benefit from an exposure to key principles of data collection and management. This is the goal of this book: to provide a basic and conceptual introduction to the collection and management of data for research.

The book is organized into four parts. Part 1 (Chapters 1 through 4) introduces the topic of data management through recent events focusing on sharing and reuse of data, and through an analysis of stories of data gone awry. Chapters 2 and 3 define data from multiple perspectives and provide methods for documenting a clear and complete definition of data for a research study. Chapter 4 focuses on data management planning including the compilation of the data management plan itself as well as the documentation of data collection and processing in the data management plan.

Part 2 (Chapters 5 through 10) covers the fundamentals of data observation, measurement, recording, and processing. Chapter 5 introduces three data milestones that determine whether, and if so how, data might be cleaned. Chapter 6 describes the characteristics of data observation and recording by humans and machines, including common sources of error with each. Chapter 7 describes the components of good documentation practices for research. Chapters 8 and 9 cover the many options for getting data into electronic format (Chapter 8) and logical ways to structure data during collection and storage (Chapter 9). Chapter 10 introduces common operations performed on data, collectively called data processing, to clean, code, integrate, and otherwise prepare data for analysis.

Today, most research involves multiple people, and sometimes many employees and large teams. Part 3 (Chapters 11 through 14) covers the management aspects of data management. Chapter 11 introduces data flow and workflow and the partitioning of tasks between humans and machines, called automation. Chapter 12 covers the complex task of identifying software needs and available software, and selecting the optimal software for data collection and management. Chapter 13 focuses holistically on the things that are needed to ensure that data quality is consistent and capable of supporting the intended analysis. Chapter 14 focuses on estimating the volume of data collection and management work, the costs associated with it, and the time over which it is likely to occur.

Part 4 provides a brief overview of data security, sharing, and archival. Chapters 15 through 17 are purposefully short in places and consist of introductory material and pointers to other resources where such resources are robust or vary by organization.

This book may be used as a primer, as a reference, or as a text. The end-of-chapter exercises are included for self-assessment as well as to support the use as a text in graduate-level research method courses.

Author

Dr. Meredith Nahm Zozus joined the University of Arkansas for Medical Sciences, Little Rock, Arkansas after an 18 years of career at the Duke University, Durham, North Carolina, where prior to joining the faculty, she served as the director of clinical data management at the Duke Clinical Research Institute, the associate director for clinical research informatics in the Duke Translational Medicine Institute, the associate director for academic programs in the Duke Center for Health Informatics.

Her career and research have focused on data quality in health care and health-related research, in particular, data collection and management methods in research. Dr. Zozus has contributed to multiple national and international data standards toward improving the consistency of data collected in health care and health-related research. Her more than two decades in research data management have been spent largely in collection and management of data for clinical studies and she continues to work with research organizations to design quality systems and infrastructure for data collection and management. Dr. Zozus has multiple federally funded grants and contracts probing questions in data quality on topics such as data standardization, data collection and management, and data quality assessment and control. In addition to published research findings, her work includes six national and international data standards and editorship of the international practice standard for management of data for clinical studies.

Dr. Zozus is a seasoned educator and has developed and taught courses across the instructional spectrum including skills-based organizational training, professional development, review courses for professional certification, graduate courses, and contributions to national curricula. She earned her masters in nuclear engineering from North Carolina State University, Raleigh, North Carolina, and her PhD in health informatics from the University of Texas School of Biomedical Informatics at Houston.

1

Collecting and Managing Research Data

Introduction

This chapter describes the scope and context of this book—principle-driven methods for the collection and management of research data. The importance of data to the research process is emphasized and illustrated through a series of true stories of *data gone awry*. These stories are discussed and analyzed to draw attention to common causes of data problems and the impact of those problems, such as the inability to use data, retracted manuscripts, jeopardized credibility, and wrong conclusions. This chapter closes with the presentation of the components of collecting and managing research data that outline the organization of the book. Chapter 1 introduces the reader to the importance of data collection and management to science and provides organizing frameworks.

Topics

- Importance of data to science
- Stories of data gone awry and analysis
- Quality system approach applied to data collection and management
- Determinates of the rigor with which data are managed
- Frameworks for thinking about managing research data

Data and Science

Data form the basic building blocks for all scientific inquiries. Research draws conclusions from analysis of recorded observations. *Data management*, the process by which *observations* including *measurements* are defined, documented, collected, and subsequently processed, is an essential part of almost every research endeavor. Data management and research are inextricably linked. The *accuracy* and *validity* of data have a direct effect on the conclusions drawn from them. As such, many research data management practices emanate directly from the necessity of research *reproducibility*, and *replicability* for example, those things that are necessary to define data, prevent *bias*, and assure *consistency*. The methods used to formulate, obtain, *handle*, transfer, and archive data collected for a study stem directly from the principles of research reproducibility. For example, *traceability*—a fundamental precept of data management is that raw data can be reconstructed from the file(s) used for the analysis and study documentation and vice versa. In other words, the data must *speak for itself*. Research data management practices have evolved from, reflect and support, these principles. These principles apply to research across all disciplines.

In a 1999 report, the Institute of Medicine (IOM) defined quality data as "data strong enough to support conclusions and interpretations equivalent to those derived from error-free data" (Davis 1999). Here, the minimum standard is the only standard that data collected for research purposes must be of sufficient quality to support the conclusions drawn from them. The level of quality meeting this minimum standard will differ from study to study and depends on the planned analysis. Thus, the question of how good is good enough (1) needs to be answered before data collection and (2) is inherently a statistical question emphasizing the importance of statistical involvement in data management planning. The remainder of this book covers the following:

1. How to define, document, collect, and process data such that the resulting data quality is consistent, predictable, and appropriate for the planned analysis.
2. How to document data and data handling to support traceability, reuse, reproducibility, and replicability.

Data management has been defined from an information technology perspective as the "function that develops and executes plans, policies, practices and projects that acquire, control, protect, deliver and enhance the value of data and information" (Mosley 2008). In a research context, this translates to data collection, processing, storage, sharing and archiving. In research, *data management* covers handling of data from their origination to final archiving or disposal and the *data life cycle* has three phases: (1) the origination phase during which data are first collected, (2) the active phase during which data are accumulating and changing, and (3) the inactive phase during which data are no longer expected to accumulate or change (Figure 1.1). This book focuses on data in origination and active phases because decisions and activities that occur in these phases most directly impact the fitness of data for a particular use, and after the origination and the active phase have passed, opportunities to improve the quality of data are slim. Other resources

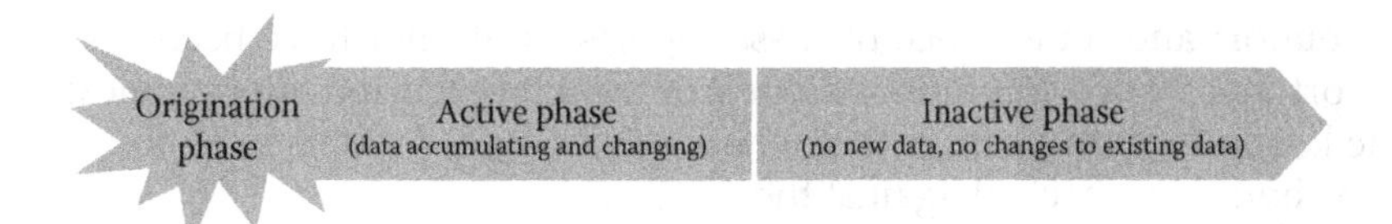

FIGURE 1.1
Phases of data management.

focus on data preservation and archival activities that occur in the inactive phase (Eynden et al. 2011, Keralis et al. 2013).

The FAIR Data Principles (Wilkinson 2016) state that data from research should be Findable, Accessible, Interoperable and Reusable (FAIR). These principles impact each phase of research data management. While Findable is achieved mainly through actions taken in the inactive phase, the foundations for Accessible, Interoperable and Reusable are created by decisions made during the planning stage, and actions taken during data origination and the active data management phase.

Data Gone Awry

Given the importance of good data to science, it is hard to imagine that lapses occur in research. However, problems clearly occur. The review by Fang, Steen, and Casadevall reported that of the 2047 papers listed as retracted in the online database PubMed, 21.3% were attributable to error, whereas 67.4% of retractions were attributable to misconduct, including fraud or suspected fraud (43.4%), duplicate publication (14.2%), and plagiarism (9.8%) (Fang 2012). These are only the known problems significant enough to prompt retraction. The following real stories include retractions and other instances of data problems in research. Each scenario given below discusses the causes of the problems, what happened, why, how it was detected, what (if anything) was done to correct it, and the impact on the research. Analyses of situations such as these illustrate common problems and can inform practice.

Case 1: False Alarm

A study published in the journal, *Analytical Methods,* reporting development and demonstration of a monitoring device for formaldehyde has earned an expression of concern and subsequent retraction for unreliable data (Zilberstein 2016, Hughes 2016). The cause of the data discrepancy was identified through subsequent work, and it was agreed that the levels of formaldehyde reported in the paper could not have been present (Shea 2016).

By way of analysis, the published statement reports that the levels originally reported in the paper would have caused significant physical discomfort and, in absence of these reports, could not have been present. Unfortunately this common sense test of the data occurred after publication. The lesson from this case is that where valid ranges for data are known, the data should be checked against them.

Case 2: Case of the Missing Eights

A large epidemiologic study including 1.2 million questionnaires was conducted in the early 1980s. The data as described by Stellman (1989) were hand entered and verified, after which extensive range and logic checks were run. On the food frequency questionnaire for the study, participants were asked the number of days per week they ate each of 28 different foods. While examining patterns of missing data, the investigators discovered that although more than 6000 participants left no items on the questionnaire blank, and more than 2000 left one item blank, no participants left exactly eight items blank (about 1100 were expected based on the distribution) and the same for 18 responses (about 250 were expected based on the distribution). The observed situation was extremely unlikely. The research team systematically pulled files to identify forms with 8 or 18 missing items to trace how the forms could have been miscounted in this way. When they did, a problem in the computer programming was identified.

By way of analysis, suspicious results were noted when the research team reviewed aggregate descriptive statistics, in this case, the distribution of missing questionnaire items. As the review was done while the project was ongoing, the problem was detected before it caused problems in the analysis. As it was a programming problem rather than something wrong in the underlying data, once the computer program was corrected and tested, the problem was resolved. There are two lessons here: (1) test and validate all computer programming and (2) look at data as early as possible in the process and on an ongoing basis throughout the project.

Case 3: Unnoticed Outliers

A 2008 paper (Gethin et al. 2008) on the use of honey to promote wound healing was retracted (no author 2014) after the journal realized that an outlier had skewed the data analysis. One patient had a much larger wound (61 cm^2) versus the other patients in the study whose wounds ranged from 0.9 to 22 cm^2. Removing the patient with the large wound from the analysis changed the conclusions of the study. The lead investigator in a response to the retraction stated, "I should have checked graphically and statistically for an outlier of this sort before running the regression, and my failing to do so led the paper to an incorrect conclusion" (Ferguson 2015a). The investigator further stated that the error came to light during reanalysis of the data in response to a query about a different aspect of the analysis (Ferguson 2015a).

In the analysis of this case, outlier in the data was not detected before analysis and publication. The outlier was only detected later upon reanalysis. As the data problem was detected after publication, correction could only be accomplished by retraction and republication. The problem could have been prevented by things such as incorporating bias prevention into inclusion/exclusion criteria or by designing and running checks for data problems. The main lesson in this case is that data should be screened for outliers and other sources of bias as early in the project as possible and ongoing throughout the project.

Case 4: Data Leaks

In 2008, the Department of Health and Human Services publically apologized after a laptop with social security numbers was stolen from a National Institutes of Health (NIH) employee's car (Weiss and Nakashima 2008). The laptop contained data from more than 1000 patients enrolled in a clinical trial. The data on the laptop were not encrypted. A few weeks later, a similar apology was made when a surplus file cabinet from a state mental health facility was sold with patient files in it (Bonner 2008). The latter problem was brought to light when the buyer of the cabinet reported his discovery of the files to the state. The seller of the file cabinet, reported that the center had moved several times and that *everyone connected to the files* were no longer there (Bonner 2008).

By way of analysis, in both cases, sensitive yet unprotected data were left accessible to others. Unfortunately, the problems came to light after inappropriate disclosure, after which little can be done to rectify the situation. The lesson here is that sensitive data, such as personal data, proprietary, competitive, or confidential information, should be protected from accidental disclosure.

Case 5: Surprisingly Consistent Responses

A colleague had run several focus groups and had recordings of each session. She hired a student research assistant to type the transcripts so that the data could be loaded into qualitative analysis software and coded for subsequent analysis. As the investigator started working with the data, she noticed that the first several transcripts seemed too similar. Upon investigation, she discovered that the research assistant had repeatedly cut and pasted the data from the first transcript instead of typing transcripts of each recording. Fortunately, the problem was found before analysis and publication; however, the situation was costly for the project in terms of rework and elapsed time.

By way of analysis, the data problem was created by misconduct of a research assistant. The root cause, however, was lack of or delayed oversight by the principal investigator. The lesson here is clear: When data are processed by others, *a priori* training and ongoing oversight are required.

Case 6: What Data?

The 2011 *European Journal of Cancer* paper titled "Expression of a truncated Hmga1b gene induces gigantism, lipomatosis and B-cell lymphomas in mice" (Fedele 2011) has been retracted (Fedele 2015). The story as reported by the retraction, watch (Oransky 2015) stated that a reader contacted the editors of the journal regarding possible duplications in two figures. When contacted, the authors were unable to provide the editors with the data used to create the figures and the journal retracted the paper in response. The corresponding author disagreed with the retraction, claiming that the source data files were "lost in the transfer of the laboratory in 2003" (Oransky 2015).

The analysis in this case is tough; the only available fact is that the data supporting the figure could not be provided. The lesson here is that data supporting publications should be archived in a manner that are durable over time.

Case 7: Different Data

Two articles involving the same researcher were retracted based on unreliable data and missing data. In 2013, after publication of the in press version of the paper titled "Early life ozone exposure induces adult murine airway hyperresponsiveness through effects on airway permeability" (Auten 2013), the authors became aware that the primary data were inconsistent with the machine-generated raw data. An earlier 2011 article, "Gastrin-releasing peptide blockade as a broad-spectrum anti-inflammatory therapy for asthma" (Zhou et al. 2011), was retracted on March 30, 2015, for the same stated reason. The retraction note states, "primary data used to calculate the in vivo pulmonary mechanics results were inconsistent with the machine-generated raw data," and *most primary data* from several other experiments were missing.

From the retraction notice, we know that processed data could not be corroborated by the raw data and that data were missing. Although details are not available, it is clear that data are not traceable back to the source data or vice versa. The lesson here is that data should be accompanied by documentation of how the data were processed, and such documentation should be in sufficient detail to enable the reproduction of the analysis from raw data.

Despite the obvious importance of sound data management practices in research, investigators and research teams sometimes fall short. The cases discussed so far demonstrate the instances of lack of training, oversight, infrastructure, documentation, security, and data checking. Good data management practices can detect these problems early or prevent them altogether. The principles covered in this book support investigators and research teams in creating and executing comprehensive plans to prospectively assure the quality of research data and to monitor the quality throughout the study.

Lapses in data management occur along a continuum that includes misunderstandings and accidental error on one end, with deliberate fraud at the opposite extreme. Lapses all along the continuum have the potential to adversely impact the conclusions drawn from scientific data and the integrity of the research. In this text, we concern ourselves with preventing nonfraudulent lapses and, where prevention is not possible, detecting and correcting them (Chapters 2 through 16). In large studies or other situations where multiple people or those other than the investigator are involved in the collection or processing of data, investigators and research teams also need to be concerned with oversight and early detection of problems (Chapter 13).

Data Management Quality System

Preventing data problems and achieving a desired level of data quality should not be a mystery; in fact, the concepts behind doing so should be known to every investigator and research team. The purpose of this book is to demystify and clearly communicate the principles behind achieving a desired level of data quality. The main concepts can be summarized in four words: *Assure then make sure* (C. Albright no date). The two concepts, assuring and making sure, are referred to as a quality system or quality management system. In quality parlance, assuring and then making sure refers to *quality assurance* and *quality control*. Although many use the two terms interchangeably, they are two distinct concepts. Quality assurance is generally thought of as activities performed in advance, such as standardizing processes, hiring qualified staff, and providing appropriate training. Quality assurance establishes the infrastructure and processes to produce the desired quality. Quality control, however, is generally thought of as ongoing activities to measure and manage quality. Quality control occurs during a process or study and serves as a feedback loop testing and reporting the quality so that corrections can be made when needed. Quality assurance includes plans for quality control. Thus, quality assurance is the *assuring* and quality control is the *making sure*. It has been shown for one common data collection process, abstraction of data from health records, that there are more than 300 factors that impact accuracy of the data (Zozus et al. 2015). The number may vary for different data collection and management processes. However, a large number of factors impacting accuracy and the presence of manual processes means that *a priori* methods, though necessary, are not sufficient, and that quality control is needed.

Quality assurance includes understanding quality requirements and those things put in place *a priori* to achieve them. Quality assurance can be broken down into understanding requirements, technical controls, procedural controls, and managerial controls. *Technical controls* are automated constraints that enforce processes, for example, a data system that automatically saves the user's identity with each entry a user makes and which can not be overridden. Another example of a technical control is the use of bar-coded labels and a bar code reader for data entry to assure that samples and their results are automatically associated with the correct identifying information—the automation eliminates the possibility of data entry error associating a sample with the wrong experimental unit. *Procedural controls* are softer than technical controls. Procedural controls are documented processes for doing things, for example, a laboratory protocol stating the ingredients and steps involved in mixing reagents. Procedural controls are often written and referred to as protocols, manuals, or standard operating procedures. Procedural controls are considered *softer* than technical controls because humans can choose not to follow or can alter procedural controls, and tracking these occurrences is often a manual process subject to the same conditions. However, technical controls enforce procedures, and where overrides are allowed, they are usually logged, that is, documented, and thus can be tracked. Managerial controls come into play because humans are involved in some capacity in most processes. *Managerial controls* are those things dictated by an organization to control things such as roles, responsibilities, oversight, and access. For example, a position description describes tasks that an individual in a job performs and the position description of the manager describes tasks associated with the oversight of performance. Other examples of managerial controls include policies for removal of computer access when a person leaves the organization and training requirements for jobs. These four things—understanding requirements, technical controls, procedural controls, and managerial controls should be in place before work starts and are major components of quality assurance.

In contrast to quality assurance, quality control is performed during processes and often on an ongoing basis. For example, in a factory, quality control involves inspections of a product as it is produced and usually at key stages during the production. Quality control detects and quantifies defects during production. The semetrics signal when the desired quality level is in jeopardy or is not achieved and, in his or her way, provides feedback that is used to adjust the process or process inputs so that the desired level and consistency of quality is contained. The main components of quality control processes include measurement or other assessment, comparison to acceptance criteria, or standards for process consistency such as limits and rules for *run charts*, and use of the measures or assessments to control the process. Quality control may be automated or semiautomated through technical controls or may occur as a manual process dictated by procedural controls. Together, quality assurance and quality control are the major tools through which investigators and research teams can ensure that data are of appropriate quality. The principles in this book enable investigators and research teams to design and create technical, procedural, managerial, and quality controls for research data collection and processing. Chapter 13 covers each of these components of the quality system for managing research data and provides a framework that can be applied to any research study (Figure 1.2).

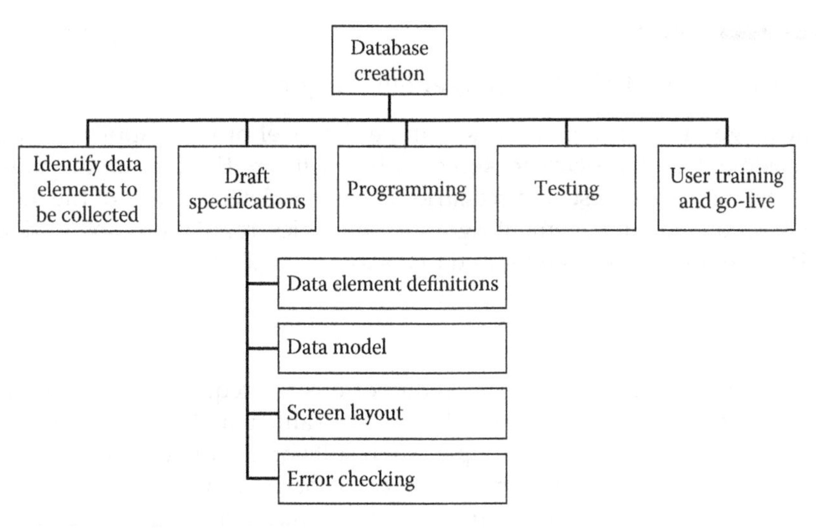

FIGURE 1.2
WBS example.

Data Management's Minimum Standards

The minimum standards for data management are data quality and traceability. These standards follow from the first principles. Data that are unable to support research conclusions are of no value. Similarly, data that are themselves not reproducible, that is, not able to be recreated from raw data to analysis dataset and vice versa, will not stand up to inquiry. For example, journal articles are retracted when the data do not support the conclusions and when analyses cannot be recreated (Ferguson 2015a, Ferguson 2015b, Ferguson 2014, Marcus 2015a, Marcus 2015b, Scudellari 2015). These two things—data quality and traceability—must be present in any scientific endeavor (Collins and Tabak 2014); thus, things necessary to bring them to fruition are the minimum standards for research data management.

Above and beyond these, however, one approach does not fit all, making research data management a profession rather than a skilled trade. In fact, the most critical responsibility of the researcher and professional data manager is matching the data management approach to the research scenario. Even within one program of research, laboratory, or organization, different studies often require different approaches and may require treatment beyond the minimum standards for quality and traceability. Throughout the text, the methodology described is necessary and sufficient for assuring that data quality is appropriate and that traceability exists. Methods over and above these such as those required by federal regulations, by certain research sponsors, or for supporting patent claims will be differentiated.

Determinates of Data Management Rigor

Knowledge of the things that determine the level of rigor applied to data management is necessary to successfully *right size* the data management approach to a given research scenario. The major determinants of the necessary level of data management rigor include ethics, regulations, institutional policies, funder requirements, and plans for data use or reuse.

Ethics

In research using human subjects, equipoise is the requirement for balance between benefit and cost to the subjects. The same may be said for communities and populations; for example, the results of a study of environmental conditions in a community may impact property values, or a study that showing increased crime rates in a population may draw stigma to that population. Situations like these pose risks and benefits. For a study that will result in increased risk to humans, there should be similar opportunity for benefit to accrue to the population from which the subjects were recruited. A study that is not at equipoise is unethical. If the data from a study involving human risk or animal suffering are not of sufficient quality to support research conclusions, then the study likewise is not equipoise. Data that are not capable of supporting the conclusions drawn from the research jeopardize, if not altogether, preclude the opportunity for benefit from the research. Although this is not a text on research ethics, the impact of data quality on equipoise bears emphasis.

Regulations

Aside from the fact that compliance with existing regulations is necessary, compliance and plans for compliance increase the perceived level of rigor of research. Regulations applicable to research come in many varieties and from different government agencies, for example, some regulations apply to materials used in research, such as the Nuclear Regulatory Commission, the Occupational Safety and Health Administration, the Environmental Protection Agency, and the Food and Drug Administration (FDA), whereas others apply to the type or intent of the research, such as the Common Rule that applies to the United States federally funded human subjects research (and all human subjects research run at institutions accepting federal funds), and FDA regulations applicable to research submitted to the FDA for marketing authorization of new therapeutics. Many other countries have corresponding regulatory authorities. Even though the goal of the research may be purely academic and for the purpose of knowledge generation, regulations may apply. Compliance is the responsibility of the investigator, and publication may depend on having complied as is the case in human subjects research, where most journals now require a statement that the research involving human subjects or animals was approved by the appropriate ethics committee. Further, regulations include both federal and state-level regulations; either or both may apply to a given study. A general text such as this cannot address regulation and guidance across all disciplines and regions of the world. The best way to identify regulations that apply to a research project is to seek advice from an organizational regulatory compliance office or from experienced investigators running similar studies. The best way to learn about the regulations is to simply read them and ask questions of institutional compliance personnel or the regulators. Such consult is especially advised where a researcher has questions about applicability of regulations to a project or about compliance options. Often, regulations are supported by guidelines, also referred to as regulatory guidance documents. The important distinction is that regulations are requirements, whereas guidelines are suggestions for how to comply—there may be more than one way to comply and the best way to find out if a proposed method is compliant is to discuss it with the appropriate regulatory authorities.

In addition to regulation and guidance, large organizations working in regulated areas usually employ a regulatory compliance specialist or have regulatory compliance groups. Usually, these groups are charged with interpreting the regulations for the organization and creating organizational policy and procedures to assure compliance. Interpretation of the regulations at one organization may be more or less conservative than at another. For example, document retention periods at organizations vary based on the regulated area, the more conservative of state and federal regulations, and organizational interpretation of the documents to which the regulated retention periods apply. Thus, investigators need to be familiar with organizational policy and procedures for regulated activities as well as the regulations themselves.

Regulations affect the level of rigor of research through requirements. For example, electronic data submitted to the FDA are subject to Title 21 CFR Part 11 that requires system validation as a part of demonstrating that electronic signatures are as good as signatures on paper. To demonstrate compliance to 21 CFR Part 11, documentation of system testing is required. Such documentation usually includes stated functions that the information system should perform, descriptions of tests for each function, and proof of testing. Validation documentation for a data processing system could reach page numbers in the thousands. However, data processing systems for research not subject to the regulation should of course be tested, but there would be no requirement for such extensive documentation of the testing.

Institutional Policies

Institutional policies and procedures are influenced by more than just regulation and organizational interpretation of regulation. With respect to data, institutional policy is often influenced by things such as organizational risk tolerance, perceived best practice, patent considerations, and local logistics. These, of course, differ from organization to organization. Data-related organizational policy and procedures at academic institutions are often located in the faculty handbook, or with information about institutional resources for investigators. Industry, however, may document such requirements as organizational policy or procedure. Organizational policies and procedures may also be adopted from among those created by consortia and other collaborative efforts.

Institutional policies and procedures may also affect the level of rigor of data-related activities. For example, institutional records retention policies often dictate the retention period, the items retained, the formats of the retained records, and where retained records may be stored. Thus, a records retention plan for identical studies may vary widely depending on the institution at which the study was conducted.

Funder Requirements

In addition to regulations and institutional policy, research sponsors are increasingly stating data-related requirements in funding solicitations, notice of grant awards, and contracts. Thus, investigators have a third place to look for requirements that may impact the level of rigor required. Copies of funder-specific requirements will usually be stated upfront in the solicitation for applications and can also be found on funder websites. Data-related requirements for contracts may be more specific and directive than those for grants. Data-related requirements of funders themselves are usually required of their contractors. For example, federal agencies in the United States are themselves subject to the Federal Information Security Management Act and the Privacy Act—obligations for compliance with these are passed along to contractors performing covered tasks. Research sponsors may contractually require that service organizations conducting research on their behalf to comply with the sponsor's procedures or other procedures. These procedures may conflict with the contractor's internal procedures; thus, awareness of sponsor requirements is critical.

Research sponsors also affect the level of rigor through requirements of funded researchers. For example, the National Science Foundation requires a two-page data management plan to be submitted as part of the application, whereas research funded by the NIH does not. Creating and maintaining a formal data management plan (Chapter 4) is clearly a level of rigor above absence of such a plan. Foundations and industry sponsors of research often have their own requirements and expectations of funded researchers.

Plans for Data Use or Reuse

Whereas the aforementioned considerations were external to the investigator and study, plans for data use or reuse may be investigator and study specific.

Plans for data use or reuse significantly affect the amount of data definition required to support the data. Although a minimum level of data definition documentation should always be maintained, in a study where the data will be used only for the planned analysis, and no further use of the data is foreseen, the study team may not need to create extensive documentation. Where data use by those other than the initial study team such as in the case of data sharing, pooled analyses, or data use far out in the future, more extensive documentation of the data definition and format will be required. Where not required by funders or institutions, investigators make the decision regarding the extent to which data will be documented. Many secondary uses of research data such as new collaborations that develop are not known at the onset, and each investigator must decide to what extent they will prospectively allocate resources to documenting data for reuse by others. The specifics of data definition are covered in Chapters 3 and 4 as are minimum standards and situations where more extensive documentation is needed.

Frameworks for Thinking about Managing Research Data

There are several perspectives from which data management can be described. The first is that of project management, that is, planning, executing, and closing out the data-related parts of a study. A second perspective involves a core set of steps followed in most research endeavors to collect and process data. The third perspective involves thinking about data management as a system. The fourth perspective involves thinking about data management in terms of the amount of control that the researcher has over the original data collection. Each vantage point highlights the important aspects of data management.

Collection and processing of research data are indeed a project; thus, many think of data management in terms of tasks that fall within phases of a project, such as planning, executing, and closing out the data-related parts of a study. Some tasks performed in data collection and processing naturally align with different stages of a project life cycle, for example, deciding and defining the data to be collected occurs at the planning stage of the project. Quality assurance and quality control tasks also fall neatly into planning and execution stages of a project, respectively. Alignment of work tasks with stages of a project is not new. The Project Management Institute and most project management software approach project management through a work breakdown structure (WBS) for a project. A WBS is a hierarchical tree of the tasks to be performed; for example, for the high-level task of database creation, low-level tasks in the WBS may include specification, programming, testing, and user acceptance testing. An example of WBS is shown in Figure 1.2. The detailed WBS is different for each project; for example, if we were creating a

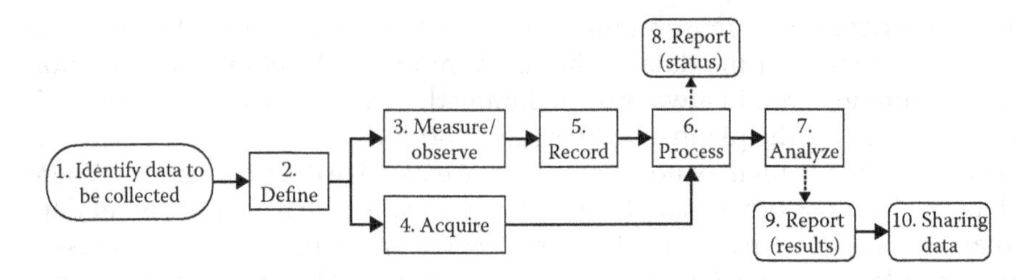

FIGURE 1.3
Research data management process.

detailed WBS for a study that collected telemetry from a migratory species as well as air, ground, and water temperature along the migration path, we might add tasks for selecting, obtaining, testing, and calibrating the devices as well as a task for the fieldwork necessary to place the devices.

The high-level steps for collecting and managing data are common across almost all research settings. Figure 1.3 shows the set of steps comprising the data-related parts of the research process. These steps are described at a general level so that they can be applied to any project. They can be used to start a WBS. Used in this way, such generic tasks would be the high-level tasks in a project WBS. Generic data collection and processing steps include (1) identifying data to be collected, (2) defining data elements, (3) observing or measuring values, or (4) acquiring data by other means as in case of secondary data use, (5) recording those observations and measurements, (6) processing data to render them in electronic form if not in electronic format already and prepare them for analysis, and (7) analyzing data. While the research is ongoing, data may be used to (8) manage or oversee the project via programmed alerts and status reports. After the analysis is completed, (9) results are reported, and (10) the data may be shared with others. Good data management principles and practices in each of these steps assure that data support the integrity of the scientific inquiry.

The third perspective from which to consider data management is that of a system. A system has inputs and outputs. The outputs or measures thereof can be used to control the system. Common inputs are listed in Figure 1.4. The output used to control the system are the metrics from data processing, for example data processing rate or the expected versus actual amount of data processed, as well as measures of data quality.

Another framework that is useful for approaching data management for a study is that of the investigator's control over the original observation or measurement and recording of the data. Although the amount of control an investigator has over the original collection of the data is in reality a continuum, breaking it down into two parts—one in which the investigator has some control and another part in which the investigator has no control (Figure 1.5)—is often helpful. It is often the case that any given study contains a mixture of these parts, that is, data from one source are under the control of

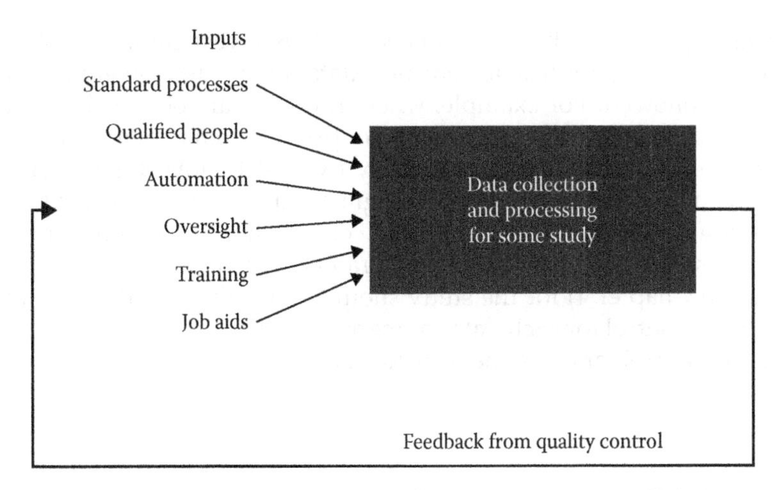

FIGURE 1.4
Data management as a system.

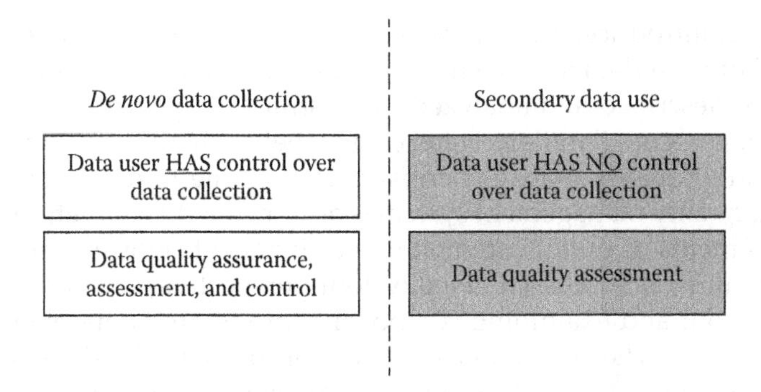

FIGURE 1.5
Investigator control and data management approach.

the investigator—for example, the investigator places and collects data from devices for the study, whereas data from another source are not—for example, the investigator uses national weather service data for air, ground, and water temperature. Thus, application of this framework is necessarily by data source.

Where the investigator has control over the data collection, full advantage should be taken of the opportunity to apply quality assurance, that is, to put procedures and controls in place that assure that the data are of acceptable quality and are traceable, including plans for quality control. Chapters 2 through 12 cover this. Where the investigator has no control over the initial data collection, the investigator is left with only the available documentation and his or her own assessment of the ability of the data to support the planned research. These two ends of the continuum dictate what should be included in a data management plan for a given data source. For data sources where the investigator has some or total control, a quality assurance and control paradigm should be followed where investigator control exists. Where the investigator has no control, a data quality assessment paradigm should be followed. For example, where raw data are obtained from elsewhere (where the investigator has no control over data collection) but are processed in the investigator laboratory (where the investigator has control), a data quality assessment paradigm should be followed for the raw data, and a data quality assurance and control should be followed for the data processed in the investigator laboratory. The data management plan (covered in Chapter 4) for the study should explicitly state the level of the investigator control for each data source and the corresponding data quality assurance, control, or assessment steps taken.

Summary

This chapter introduced the concept of research data management. The relationship between data and science were discussed. Examples of data gone awry were described and analyzed to illustrate key principles of research data management. Parallels were drawn between these principles and application of quality management system thinking to data management. Because a quality management system is designed with a target, some quality requirements in mind, common determinants of rigor in research data management were discussed. Finally, four perspectives or frameworks for thinking about and examining components of research data management were presented. The remainder of this book presents fundamental principles and methods for the collection and management of data for research.

Exercises

1. In the following scenario, identify the root cause of the problem, describe the impact, and discuss the approaches that could have prevented the problem.

   Data for a study of factors contributing to employee turnover were single entered. During the analysis, the investigator reviewed the statistical output and knew something was wrong because several factors known to be unimportant scored high. When the investigator and statistician investigated the problem, they discovered that a data entry operator had consistently transposed yes and no answers from the data collection form entering yes as no and vice versa.

Correct answer: Options for the root cause: (1) the transposition by the data entry operator and (2) lack of quality control or some type of review before analysis. Options for the impact: (1) rework or reentry, (2) manual review to identify the transposed items and their correction, and (3) rerunning of the analysis. Approaches that could have prevented the problem: (1) use of double rather than single data entry and (2) quality control of a random sample of forms.

2. In the following scenario, identify the root cause of the problem, describe the impact, and discuss the approaches that could have prevented the problem.

   An investigator obtained biological survey data for amphibian populations in a three-county region. He used the relative census numbers and species types to estimate the amphibian contribution to food sources for local fish populations. After publication, a second investigator did a separate study using their own species census survey and expected results conforming the initial paper. Unfortunately, the results of the first paper were refuted by the second paper. The author of the second paper investigated the survey data used by the initial investigator and discovered that the survey included only adult populations of amphibians, significantly underestimating the available amphibian food sources for local fish.

   Correct answer: Options for the root cause: (1) lack of data documentation in the initial survey data and (2) the first investigator used data without fully understanding it (lack of data quality assessment). Options for the impact: (1) retraction or publishing a clarification of the first paper, (2) extra work caused for the second investigator who identified the problem, and (3) blemished career of the first investigator. Approaches that could have prevented the problem: (1) more complete characterization of the survey data by the sharer and (2) better investigation of the survey data by the first investigator.

3. In the following scenario, identify the root cause of the problem, describe the impact, and discuss the approaches that could have prevented the problem.

   John C., who is a postdoc, finished his research on characterizing material erosion from different surfaces considered for the interior surface of industrial laundry vats and left the university to pursue a career in the chemical industry. Five years later after publishing the work with the head of the laboratory in which he worked, questions arose about the results because the best performing surface was significantly more expensive, and a thorough engineer considering the best performing surface for a similar application wanted more information. The laboratory head who was also the corresponding author of the publication did not have a copy of the data. When he contacted the postdoc, he said that he disposed of the disks when he moved to a new house 2 years prior.

Correct answer: Options for the root cause: (1) lack of data documentation and (2) lack of data retention policies and enforcement in the lab. Options for the impact: (1) retraction of the paper and (2) blemished career of the authors. Approaches that could have prevented the problem: (1) data retention policies in the laboratory, (2) periodic checking of data retention in the laboratory, and (3) data retention policy and associated audit by the university.

4. Describe a situation in your field (if applicable) where a researcher might have little or no control over initial data collection and a situation (if applicable) where a researcher has significant control over data collection. Describe the data quality assurance and control that might be applied to each.
5. An investigator plans to use data from a national survey. The data are available for download from a Website. Which of the following apply?
   a. Quality assurance
   b. Quality control
   c. Quality assessment
   d. All of the above

   Correct answer: (c) Quality assessment. In the example, the investigator has no control over the collection and processing of the data (before he or she downloads them); thus, the best the investigator can do is to assess the ability of the data to support his or her planned analysis.
6. An investigator is planning a study where he or she is using soil sample analysis from a prior lunar mission. He or she will collect new samples from the same locations on an upcoming mission and compare the two. Which of the following apply?
   a. Quality assurance
   b. Quality control
   c. Quality assessment
   d. All of the above

   Correct answer: (d) All of the above. The investigator is using already collected and analyzed data over which he or she has no control; in addition, he or she is collecting new data over which he or she does have control.
7. List regulations that are applicable to your field and describe any that cover data-related issues. Consider regulations applicable to federally funded research as well as those applicable to relevant

industry funded research. This may take some time but is incredibly important to your development as an independent investigator.

8. Locate your organization's record retention policy.
9. For funding agencies or foundations in your field, find out if they require a data management plan.
10. Explain why standard operating procedures (SOPs) are considered quality assurance.

    Correct answer: SOPs are written and approved before an activity takes place. The purpose of SOPs is to assure consistent performance and thus quality by standardizing a process.
11. For the scenario described in the first exercise, list tasks that may have benefitted from SOPs.

    Correct answer: Data entry, data quality control.
12. For the scenario described in the first exercise, describe a quality control procedure that would have likely detected the problem.

    Correct answer: For a statistically representative sample, compare the data forms to the data as entered into the database.
13. A political science research project requires administering a web-based survey to students enrolled in political science courses at a particular university. Which of the following are the tasks that would likely be included in a WBS for the study?

    a. Develop survey, administer survey, and analyze data
    b. Purchase accelerometers, calibrate accelerometers, and distribute laboratory kits
    c. Download national survey data and complete data use agreement
    d. None of the above

    Correct answer: (a) Develop survey, administer survey, and analyze data. These tasks are the most reasonable for a survey study.
14. Which of the following is an example of a quality control task?

    a. Writing and implementing standard operating procedures
    b. Developing an organizational quality policy
    c. Training new employees
    d. Performing an interim database audit

    Correct answer: (d) Performing an interim database audit. The other options are examples of quality assurance procedures.
15. Identify one or more data related retraction by searching the tab, *data issues* on Retraction Watch. Analyze the case and describe what went wrong, and what could have prevented the problem or mitigated the impact.

References

Albright C. Managing research data: a decade of Do's and Don'ts. Albright Consulting Web site. Available at: http://www.visi.com/~chaugan/carol/Managing_Data_Talk.htm (accessed June 7, 2007). No longer available.

Auten RL, Mason SN, Potts-Kant EN, Chitano P, and Foster WM. Early life ozone exposure induces adult murine airway hyperresponsiveness through effects on airway permeability. *Journal of Applied Physiology* 2013. Published 15 June 2013 Vol. 114 no. 12, 1762 doi: 10.1152/japplphysiol.01368.2012.

Bonner L. DHHS sells a file cabinet—Files and all. *Raleigh News and Observer*, November 22, 2008.

Collins FS and Tabak LA. NIH plans to enhance reproducibility. *Nature* 2014, 505:612–613.

Davis JR, Nolan VP, Woodcock J, and Estabrook RW. (eds.). Assuring data quality and validity in clinical trials for regulatory decision making. *Workshop Report. Roundtable on Research and Development of Drugs, Biologics, and Medical Devices. Division of Health Sciences Policy*, Institute of Medicine. Washington, DC: National Academy Press, 1999.

Fang FC, Steen RG, and Casadevall A. Misconduct accounts for the majority of retracted scientific publications (2012). *Proceedings of the National Academy of Sciences Early Edition*, online October 1, 2012.

Fedele M, Visone R, De Martino I et al. Expression of a truncated Hmga1b gene induces gigantism, lipomatosis and B-cell lymphomas in mice. *European Journal of Cancer* 2011; 47(3):470–478.

Fedele M, Visonea R, De Martinoa I et al. Retraction notice to expression of a truncated Hmga1b gene induces gigantism, lipomatosis and B-cell lymphomas in mice. *European Journal of Cancer* 2015; 51(6):789.

Ferguson C. JCI lymphoma paper retracted after authors can't find underlying data to explain duplicated bands. *Retraction Watch* 2014. Accessed May 13, 2015. Available from http://retractionwatch.com/2014/12/08/jci-lymphoma-paper-retracted-authors-cant-find-underlying-data-explain-duplicated-bands/#more-24133.

Ferguson C. *Sticky situation: Paper using honey for healing pulled over data issues* 2015a. Accessed May 13, 2015. Available from http://retractionwatch.com/2015/03/18/sticky-situation-paper-using-honey-for-healing-pulled-over-data-issues/#more-26346.

Ferguson C. Chem paper retracted because "a significant amount of data" was wrong. *Retraction Watch* 2015b, Accessed May 17, 2015. Available from http://retractionwatch.com/2015/02/10/chem-paper-retracted-significant-amount-data-wrong/#more-25585.

Gethin GT, Cowman S, and Conroy RM. The impact of Manuka honey dressings on the surface pH of chronic wounds. *International Wound Journal* 2008; 5(2):185–194.

Hughes P. Expression of concern: Remote sensing of formaldehyde fumes in indoor environments by P. G. Righetti et al., *Analytical Methods* 2016; 8:5884. doi: 10.1039/c6ay00976j.

Keralis Spencer DC, Stark S, Halbert M, and Moen WE. Research data management in policy and practice: The DataRes Project, In Research data management principles, practices, and prospects, Council on library and information resources (CLIR), CLIR Publication No. 160, Washington DC, November 2013.

Marcus A. Slippery slope? Data problems force retraction of landslide paper. *Retraction Watch,* 2015a. Accessed May 17, 2015, available from http://retractionwatch.com/2015/04/01/slippery-slope-data-problems-force-retraction-of-landslide-paper/#more-26882.

Marcus A. Data "irregularities" sink paper on water treatment. *Retraction watch* 2015b, Accessed May 17, 2015, available from http://retractionwatch.com/2015/03/26/data-irregularities-sink-paper-on-water-treatment/#more-26448.

Marcus A. He shoots, he...misses! Soccer injury paper gets red card for data errors. *Retraction watch,* 2015, Accessed May 17, 2015, available from http://retractionwatch.com/2015/02/05/shoots-misses-soccer-injury-paper-gets-red-card-data-errors/#more-25647.

Mosley M. (ed.). *The DAMA dictionary of data management,* 1st ed. New Jersey: Technics Publications LLC, 2008.

No author. Retraction. The impact of Manuka honey dressings on the surface pH of chronic wounds. *International Wound Journal* 2014; 11(3):342.

No author. Retraction notice to "Expression of a truncated Hmga1b gene induces gigantism, lipomatosis and B-cell lymphomas in mice" [Eur J Cancer 47 (2011) 470–478]. *European Journal of Cancer* 2015; 51(6):789.

No author. Retraction notice, *Journal of Applied Physiology* 2013; 114(12):1762. doi: 10.1152/japplphysiol.01368.2012.

No author. Retraction notice for Zhou et al. Gastrin-releasing peptide blockade as a broad-spectrum anti-inflammatory therapy for asthma. doi: 10.1073/pnas.1504672112.

No author. Retraction of early life ozone exposure induces adult murine airway hyperresponsiveness through effects on airway permeability. *Journal of Applied Physiology* 2013; 114:1762.

No author. Retraction for Zhou et al. Gastrin-releasing peptide blockade as a broad-spectrum anti-inflammatory therapy for asthma. *Proceedings of the National Academy of Sciences of the United States of America* 2015; 112(14):E1813.

Oransky I. Cancer researcher under investigation in Italy notches eighth retraction. *Retraction Watch* 2015. Accessed May 13, 2015, available from http://retractionwatch.com/2015/04/06/cancer-researcher-under-investigation-in-italy-notches-eighth-retraction/#more-26989.

Scudellari M. "Unreliable" data suffocates third paper for Duke pulmonary team. *Retraction watch* 2015. Accessed May 13, 2015, available from http://retractionwatch.com/2015/05/04/unreliable-data-suffocates-third-paper-for-duke-pulmonary-team/#more-27860.

Shea CD. False Alarm: Damien Hirst's Formaldehyde Fumes Weren't Dangerous. *New York Times,* 2016.

Stellman SD. The case of the missing eights. *American Journal of Epidemiology* 1989; 120(4):857–860.

Van den Eynden V, Corti L, Woollard M, Bishop L, and Horton L. Managing and sharing data. *The UK Data Archive,* 3rd ed. University of Essex, Wivenhoe Park, Colchester Essex, 2011.

Weiss R and Nakashima E. *Stolen NIH Laptop Held Social Security Numbers. Washington Post,* 2008.

Wilkinson MD et al. *The FAIR Guiding Principles for scientific data management and stewardship.* Sci. Data 3:160018 doi: 10.1038/sdata.2016.18 (2016).

Zhou S, Potts EN, Cuttitta F, Foster WM, and Sundaya ME. Gastrin-releasing peptide blockade as a broad-spectrum anti-inflammatory therapy for asthma. *Proceedings of the National Academy of Sciences of the United States of America* 2011; 108(5):2100–2105.

Zilberstein G, Zilberstein R, Maor U, Baskin E, Zhang S, Righetti PG. Remote sensing of formaldehyde fumes in indoor environments by P. G. Righetti et al., *Analytical Methods* 2016. doi: 10.1039/c6ay00976j.

Zozus MN, Pieper C, Johnson CM, Johnson TR, Franklin A, Smith J et al. Factors affecting accuracy of data abstracted from medical records. *PLoS One* 2015; 10(10): e0138649.

2

Defining Data and Information

Introduction

Before we discuss the details of managing data, it is important to pause and explore the nature of data, for example, questions such as *what are data* and *how do they differ from information and knowledge*? Such questions lead to an understanding of the fundamental kinds of data and some aspects of data that indicate methods of collecting and managing them. As least as early as the early 1900s and even back to the time of Aristotle, people contemplated these questions; some remain in discussion even today. This chapter provides an overview of some of the key foundational ideas and principles that define data, information, and knowledge. Fundamental definitions are provided, and distinctions are drawn that deepen our understanding of that which investigators and research teams seek to collect, manage, analyze, and preserve.

Topics

- Semiotic theory
- Data as a model of the real world
- Defining data and information, including common use of the terms *data* and *information*
- Introduction to data elements
- Categories of data

Data about Things in the Real World

The ideas of Charles Sanders Peirce, an American philosopher and scientist (1839–1914), define an important aspect of data. Peirce explained a triad including an *immediate object*, that is, some object in the real world; a *representamen*, that is, some symbol representing the object; and an *interpretant*, that is, a thought or reference of the object in the mind linking the representation to the actual real-world object. (Pierce 1931a, 1931b, 1977) Ogden and Richards (1923) constrained and brought a simpler language to Pierce's ideas to create what is known today as the semiotic triangle (Figure 2.1). (Ogden and Richards 1923) Briefly, the semiotic triangle depicts the relationship between mental concepts, things, and symbols. A thing is something in the real world, for example, an object, a dream, or a song. This thing in the real world is called a referent, as in a referent to which some symbol might refer. The symbol represents the thing, but it is not the thing; for example, a symbol could be an icon, a sign, or a gesture. A mental concept is a perception, some thought or idea in *the mind's eye* of what the thing in the real world is. Relating things in the real world, symbols, and an individual's perception of the thing while at the same time maintaining that the thing, the idea, and the symbol are three different has interesting consequences. As the three are separate, a concept can be decoupled from symbols that represent it, and there may be several possible symbols to represent a real thing. There may also be different mental concepts of that same real world thing. For example, the concept *danger* has multiple representations, two of which are the Jolly Roger and the green Mr. Yuck symbols.

Building on the semiotic triangle, a data value can be thought of as a symbol. For example, the symbol 212°F represents the level of temperature. The data value 212°F may invoke a mental concept of the steam rising from boiling water, whereas the thing in the real world is the temperature of water in a beaker in someone's laboratory. In this way, the data value is a symbol representing a phenomenon in the real world. Thus, data are not the thing itself, but can be thought as merely the symbols representing the thing. Similarly, data are yet again distinctly separate from the software, databases, paper, files, or other media in which we store them—data can be moved from one storing format to another.

Following the semiotic triangle, Morris (1938) and Stamper (1991) put forward a framework, distinguishing four levels at which symbols are used.

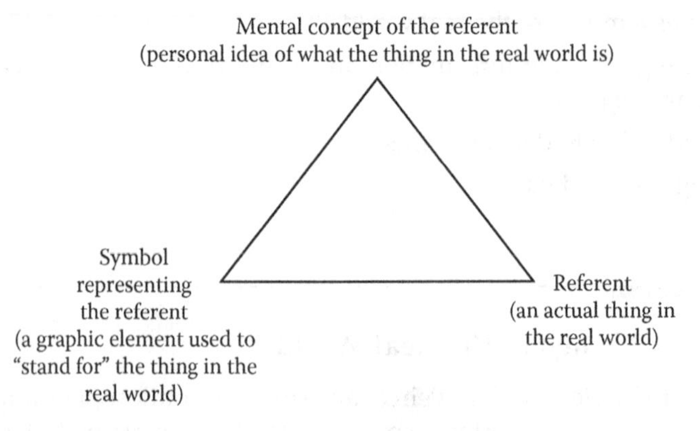

FIGURE 2.1
Data and the semiotic triangle.

The first level, *empirics,* is defined as the transmission of signs and signals. The second level, *syntactics,* is defined as the formal properties of sign systems. The third level, *semantics,* is defined as the meaning of signs, that is, what is implied about the world given that the sign exists. Finally, *pragmatics* is defined as the use of signs. This framework applies to any system of symbols, nautical flags, air traffic control signals, road signs, sign language, or the code. Some systems are quite complex, for example, whole languages, or quite simple, for example, Paul Revere's *One, if by land, and two, if by sea.* Applying the framework to Revere's lanterns, the empirics (transmission of the signs) included the use of light at night. The syntactics included one, two, or no points of light in a church steeple. The semantics of the system are quite famous, if Revere could not make the ride, there would be one lantern lit if the British were coming by land and two if by sea. The pragmatics included displaying the lanterns high in the church steeple and their interpretation as a que to action (Table 2.1).

As a second example, consider the traffic light as a sign system. In most countries, there are three colors: red, yellow, and green. Empirically, the signs are transmitted visually by different color lights. Syntactically, there are three values corresponding to the colors, and in some countries, the same color arrows are used, so syntactically three colors and two symbols (solid circle and arrow) use two channels of communication. Semantically, the red, yellow, and green mean stop, proceed with caution, and go while the arrows signal the same for protected turning traffic, respectively. Pragmatically, the sign system is automated and used to control traffic.

Data are symbols representing singular statements about things in the real world, for example, temperature or color. There are a very large number of such statements needed to completely describe anything in the real world. Individual data values themselves rarely provide enough information. For example, someone shopping for cars may have 20 of 30 attributes based on which they will compare cars, for example, the make, model, engine capacity, fuel efficiency, price, electric locks, and the presence of a sunroof. There are an even larger number of statements that differentiate one care from another, or that would be required for a factory to order the parts and materials to assemble a car. The factory likewise may need to maintain data on the manufacturing process, maintenance of machines, employees, and financial performance. Any set of data may represent a very many singular statements. Organizing them and understanding the ways in which they relate to each other become important.

TABLE 2.1

Semiotic Framework

Empirics	Transmission of signs and signals
Syntactics	Properties of the sign system
Semantics	Meaning of the signs
Pragmatics	Use of the signs

Data as a Model of the Real World

Statements about things in the real world can be thought of as a model of the real world. The industrial statistician George E.P. Box is well known for having said, "Essentially, all models are wrong, but some are useful. ... the approximate nature of the model must always be borne in mind ...." (Box and Draper 2007). A collection of data will always incompletely describe the state of the real world. A complete description would be sufficient to create a complete replication, anything else falls short. For this reason, some collections of data may support answering some questions about the real world but not others. The questions answerable by data are determined by how completely the data represent the aspects of the world about which the questions are asked.

Consider the following example: vehicle registration in the United States requires the collection of the make, model, year, owner, license plate number, and a vehicle identification number. Government driver's license data contain demographic information about licensed drivers. A researcher could not use a vehicle registration database in conjunction with the driver's license data to test a hypothesis that certain vehicle options such as sun roofs and sports suspension were more prevalent in certain demographics because neither the vehicle registration data nor the driver's license data contain statements about vehicle options such as sun roofs and sports suspension. Thus, the data (the model) are incomplete with respect to the aspects of the real world that are the topic of our question. This is commonly the case when reuse of existing data is attempted. *Reuse* of data is the use of data for purposes other than those for which the data were originally collected.

A related concept often discussed with models is the *level of abstraction (LoA)*. LoA is commonly thought of as the detail level at which the data exist. For example, a map of a country showing major cities is useful in gauging geographic proximity of the cities. However, such a high-level map does not show roads, bodies of water, bridges, or rail routes; thus, the map is not useful for navigating from one city to the next. Likewise, traveling by foot, bike, or bus requires even more detailed information, that is, less abstraction. Data at a lower LoA can be aggregated up to answer questions at a higher LoA, for example, using road or mileage information to answer the question of geographical proximity of certain cities, but not the reverse. By virtue of describing the world at the higher LoA, the first map omits the more detailed information. Thus, the lower LoA is a more detailed model. LoA also determines whether a collection of data is capable of supporting some intended use.

Defining Data

Although the term *data* has been variously defined, many of the definitions are in keeping with data as a symbol representing something in the real world. For example, the international standard ISO/IEC 11179 defines data as "a re-interpretable representation of information in a formalized manner suitable for communication, interpretation, or processing" (ISO/IEC 11179 2013). Similarly, the data management professional association defines data as "Facts represented as text, numbers, graphics, images, sound or video ..." and goes on to say that "Data is the raw material used to create information" (Mosley 2008). Popular definitions of data stray from data as a representation and tend toward how data are used: "factual information (as measurements or statistics) used as a basis for reasoning, discussion, or calculation," "information output by a sensing device or organ that includes both useful and irrelevant or redundant information and must be processed to be meaningful," and "information in numerical form that can be digitally transmitted or processed" (Merriam-Webster 2015). The popular definitions, in addition to straying from data as a representation, tend to imply that data and information are synonyms and tend to imply equivalency between data as a representation as a singular fact and some aggregate form of data such as a sum or an average. These distinctions are helpful and should not be glossed over. Useful definitions will clarify such important distinctions that inform methods for handling data.

Data and information have been given significant attention in the Information Theory literature. An extensive treatment of the evolution of and different perspectives on defining data and information are beyond the scope here.

The diaphoric definition of data (DDD) and the general definition of information (GDI) as described by Floridi (2015) are also based on this line of thinking. Diaphoric definition of data is as follows: "a datum is a putative fact regarding some difference or lack of uniformity within some context" (Floridi 2017). Note the word putative only signifies that the data are alleged to be facts, not that they are.

Such differences in the real world exist in multitudes, occur naturally, and are created. Examples such as ridges on the plates of a turtle's shell, concentric rings on a tree, and the order of nucleotides in deoxyribonucleic acid occur naturally; they are generated through natural processes. Alternately, other diaphora are created, for example, a paw print created by an animal walking through loose soil, or the dimensions of a room resulting from a construction process. Distinguishing between naturally occurring and created diaphora is probably not important to data management. However, the nature of the diaphora and the manner in which they are observed are of great importance.

Three Fundamental Kinds of Data

Some differences, from here called phenomena, are directly observable and others are not. For example, attendance at a meeting is directly observable, whereas attendee perception of the value of the meeting is not directly observable—though we might get some indication through behavior. Those phenomena that can be perceived directly by a human observer can be the presence or absence of something, for example, seeing a paw print; counts of things, for example, number of organisms encountered per square mile; or relative differences, for example, the three beds Goldilocks encountered—one that was too small, one that was too large, and one that was just right. Direct perception may also be part of a more complex process of noticing and then interpreting such as seeing a white spot on a chest X-ray and thinking that it may be lung cancer—direct observation and further interpretation should not be confused. Thus, direct perception is one way of observing a phenomenon of interest.

Measurement of physical quantities is another way of observing a phenomenon, for example, measuring the length and width of a paw print, measuring the pH of a liquid, particle count and size distribution of particulate contamination, or using a device to take a cell count. All physical quantities have associated with them a fundamental dimension. The basic physical dimensions are length, mass, time, electrical charge, temperature, luminous intensity, and angle. There are multiple systems of units for measuring physical dimensions. The major unit systems include the following: (1) the metric MKS system based on meters, kilograms, and seconds; (2) the metric CGS system based on centimeters, grams, and seconds; and (3) the English system based on feet, pounds, and seconds. The dimensionality of a physical quantity is a property of the phenomena and is independent of the unit system. For example, a runner who traverses 1 mile covers the same distance as a runner who runs 1.60934 km; distance in miles and distance in kilometers both are two expressions of the fundamental dimension length. There are many derived quantities that are combinations of the basic dimensions, for example, velocity represents the length per unit or time. Physical quantities such as force, viscosity, weight, torque, current, capacitance, resistance, and conductivity are all derived quantities based on the fundamental dimensions. The dimensions of any physical quantity can be expressed in terms of the fundamental dimensions. The dimensional system together with unit systems based on the fundamental units and the standards that define them provides a basis for measurement of all physical quantities and the comparison of measurements of physical quantities. In addition to dimensionality and units, representation of the measurements of physical quantities necessarily must match the measured phenomena. As such, the accuracy (extent to which the measurement and the true value agree) and precision (the smallest increment to which the quantity can be reliably measured given some measurement method) of measurements of physical quantities are also important.

Other phenomena are not directly observable, for example, some phenomena are directly experienced by an effected individual, for example, level of physical pain, an opinion, or recollection of a past event. In some situations, these can be ascertained by asking the experiencer directly. Sometimes, however, asking the experiencer is not possible, for example, asking an animal if it is adversely impacted by pollution. The latter is an example of a phenomenon that is not directly observable. Fortunately, some phenomena that are not directly observable may be ascertained through observing something else that is related in some known way, for example, some measure that has been validated to correspond with the phenomena of interest. This is called a *surrogate*. Some surrogates are directly perceivable, for example, in case of two species that do not coexist, seeing one indicates that the other is not present. Other surrogates are measured quantities. For example, some aspects of cardiac activity are measured as changes in current through leads attached to the chest. These leads do not allow direct measurement of cardiac activity, but instead measure current induced from the coordinated movement of ions across myocardial cell membranes. Certain features of the tracing of the current correspond to cardiac activity—a count of the number of regular peaks per minute indicates the number of times per minute the heart beats. Another example is the detection of large weapons testing across the world by picking up very low-frequency waves. Further, one that is not directly observable today may become so tomorrow through advances in technology enabling different types of measurement or through advances in understanding such as identification of surrogates.

The discussion in this section covered three fundamental kinds of data. The three fundamental kinds of data can be characterized by (1) the observeability of the phenomena of interest and (2) the observation method (perception, measurement of a physical quantity, or asking). This leaves us with direct perception, direct measurement, and asking as three fundamental ways to obtain data (Figure 2.2). A researcher should never lose sight of the fundamental kinds of data with which they are working. Some phenomena can be ascertained by multiple methods. Many phenomena can be

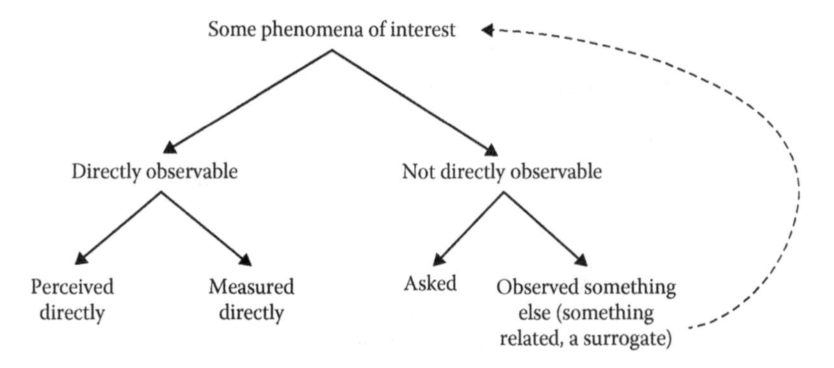

FIGURE 2.2
Fundamental kinds of data.

both directly perceived and directly (or indirectly) measured, for example, heart rates can be counted by a human for a period of time or can be measured with a heart rate monitor. A human having an asthma attack directly observed through the outward signs, measured by peak expiratory volume, or the asthma attack can be reported by the human.

Data Elements

The real-world phenomena of interest, the observation method, and the resulting data come together in a data element. A *data element* is a question–answer format pair (Figure 2.3). For example, the flow rate from a faucet measures how much water comes out of the faucet in a given period of time. We might measure the flow rate in gallons per minute by placing a 5-gallon bucket under the faucet and reading off how many gallons of water are in the bucket at the end of 1 minute. To increase the reliability of the measurement, we may use a larger bucket and measure for 5 minutes. The value that we measure from a showerhead may be 2.0 gallons per minute. Here, the measurement, 2.0 gallons per minute, is the data value. If we measure every showerhead in an apartment building, we have as many data values as we do time points at which the flow rate from each showerhead was measured. The data element provides the meaning behind the measurement. Associating the data value 2.0 with the data element tells us that 2.0 refers to a flow rate rather than the number of ducks crossing the road. The data element carries the definitional information about the data value. Here, the data element carries the definition of flow rate possibly including the measurement method; it carries the units of the measure, for example, gallons per minute and even the precision to which the value is measured (Figure 2.3). Other data elements collected with

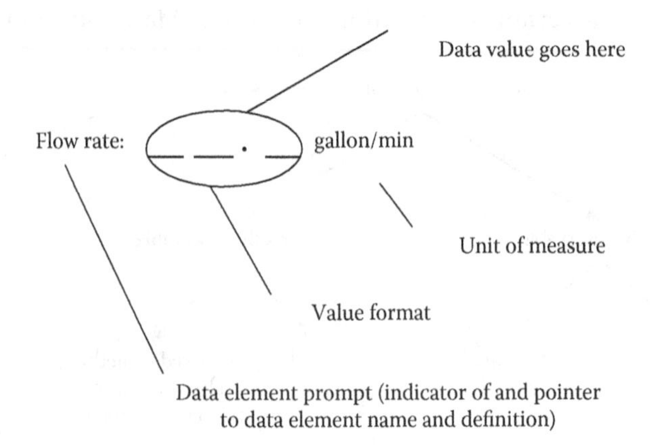

FIGURE 2.3
Anatomy of a data element.

the flow rate may carry an important contextual an information necessary for appropriate interpretation of the measurement such as the measurement date, the apartment number, the bathroom number, the floor (might be associated with a different water pressure for each floor), the time of day, and the person taking the measurement. Data elements are the atomic unit of information.

Stevens's Scales of Measurement

Data in research are usually generated by the observation of phenomena. Phenomena have an inherent and set amount of observable information. An observation method may provide the full amount of observable information or some lesser amount. Consider some phenomena such as surface temperature of a road in the summer detailed measurements taken every hour would provide a daily temperature distribution. Such a level of detail compared to the timescale of the phenomena would be considered a low LoA. However, a higher LoA such as peak and trough temperature may suffice for the research. Further, the context in which the phenomena occurred may be of importance to the research. Collecting additional contextual information such as cloud cover, rain, and whether the road was in the shade increases the scope of the observation through the addition of data elements for some measurement of cloud cover, rain, and shade. A researcher interested in a particular phenomenon may desire the full amount of observable information OR something less. A key decision that a researcher must make for each phenomenon of interest is the amount of information needed about the phenomenon of interest and how that information might be obtained. One way to assess the information content is through Stevens's scales of measurement.

Three separate and important concepts such as (1) the amount of information inherently observable in a phenomenon, (2) the amount of information desired, and (3) the amount of information resulting from an observation method can all be described in terms of Stevens's scales (Stevens 1946). Briefly, *Stevens's scales* are four exhaustive and mutually exclusive categories based on the information content of measurement. They are nominal, ordinal, interval, and ratio. The *nominal scale* applies to labels, for example, gender, hair color, country of origin, and so on. Items on a nominal scale have no order and do not have any numerical significance. Things measured on a nominal scale are qualitative; they carry no information other than a category assignment such as gender of male or female. The *ordinal scale* type also consists of labels, but additional information exists through some order, for example, small, medium, large or mild, moderate, or severe. Although they have order, things measured on an ordinal scale are also qualitative and without numerical significance. The *interval scale* type is a numerical scale. The interval scale is quantitative and the exact value is knowable. Interval scales carry additional information content over that of ordinal scales because the space between points on an interval scale is the same. Time is a good example of an interval scale, for example, the temporal distance (120 minutes) between 1 o'clock and 3 o'clock is the same as the distance between 5 o'clock and 7 o'clock. However, on an interval scale, there is no absolute zero. Thus, measurements on an interval scale can be compared to each other, but we have no way of knowing if one measurement is twice as large as another. A *ratio scale,* also numeric, carries the maximum amount of information because not only is the distance between two points on the scale the same, but there is an absolute zero, for example, temperature measured in degrees Kelvin, weight or height as given in Table 2.2.

Stevens's scales offer a way to assess and check the alignment between the scale inherent in an observed phenomenon, the scale of the desired information, and the scale associated with the observation method. For any combination of phenomena and desired information, there are often multiple measurement methods. Optimally, a researcher will use a measurement method that provides the desired information content and preserves as much of the inherent information content as possible with the given available resources. Using an observation method that delivers less information than is observable in the phenomena is an example of data reduction. For example, many clinical studies measure cholesterol due to its importance in cardiovascular outcomes. The laboratory test for cholesterol is inherently a ratio scale, and early clinical studies used measurement methods that provided the numeric values on a ratio scale. However, the desired information in many of these studies was nominal, that is, whether a person had high cholesterol or not, a *dichotomous measure.* Studies that collected the nominal value thus reduced the available information content from a ratio to a nominal scale. Later, the definition of high cholesterol changed and the old data were less valuable because the labels were no longer accurate for those values affected by the definition change. Because the data were measured at the

TABLE 2.2

Properties of Stevens's Scales

	Nominal	Ordinal	Interval	Ratio
Quantitative continuous			✓	✓
Qualitative categorical	✓	✓		
Has absolute zero				✓
Equidistant intervals			✓	✓
Can be ordered		✓	✓	✓
Values can be used as labels	✓	✓	✓	✓
Will support counts	✓	✓	✓	✓
Will support mode and median		✓	✓	✓
Will support addition and subtraction			✓	✓
Will support multiplication and division				✓

inherent scale and reduced when documented for the study, the reduction did not save any effort. However, it is sometimes the case that phenomena are measured at a lower scale than the inherent information content because the latter is less expensive, or because the additional information content is not important to the study. Urine versus plasma pregnancy tests are an example. The urine dipstick test returns an immediate result of positive or negative and is inexpensive, whereas the blood serum test takes longer and is more expensive, but measures serum human chorionic gonadotropin on a ratio scale. For many uses, the dipstick measurement is sufficient and the data reduction is considered a fair trade-off.

By providing an indication of the amount of information inherent in a phenomenon, desired by a researcher, or available through an observation method, Stevens's scales help us differentiate matches and nonmatches and make such decisions explicit.

Back to the Definition of Data and Information

Now that we have explored *data* as diaphora in terms of observability, measurement, and obtainable information, we will return briefly to the definition of data and its relation to other terms such as information and knowledge.

DDD defines data as a nonsemantic entity—recall "a datum is a putative fact regarding some difference or lack of uniformity within some context." Some have stopped here at data as diaphora without regard to meaning. For some applications such as data processing, this approach may be quite appropriate and in fact sufficient. For example, Shannon's Mathematical Theory of Communication (MCT) treats data without regard to the semantics. MCT deals with the flow of data through a channel such as a telephone line and quantifies the amount of data as well as concepts such as noise, loss, and lossless transfer. In this case, noise and loss are the two threats to accuracy. For use of data solely by machines in areas such as computer science, data storage, and signal transmission, treatment of data without regard to semantics is fine. Semantics, however, are important to collection and management of data in research.

Applications involving human use of data require that the meaning of the data be taken into account. Floridi provides for consideration of semantics through the GDI. GDI states that information equals data plus meaning and goes on to require that information consists of one or more data, that the data are well formed, and that the well formed data are meaningful. Defining information as data plus meaning also has roots in the ideas of others (Mingers 1997, 2012, Floridi 2017, Devlin 1991).

Today, research applications require both human and machine use of data. Thus, definition of data that supports both is necessary. There are two perspectives with respect to data and meaning. In an objectivist stance with respect to semantics, information is viewed as independent of the receiver, that is, they have their own semantics independently of an intelligent receiver (Floridi 2015, Davis and Olson 1985, Silver and Silver 1989, Dretske 1981). An alternate view is one of subjectivity where information is viewed as constructed by the receiver from the data by assigning meaning to the data possibly by adding the receiver's own meaning (Checkland and Holwell 1998, Lewis 1993). The latter allows the possibility that different receivers generate different information from the same data. In the objectivist approach, the meaning in *data plus meaning* is a property of and comes with the data, whereas from the subjectivist point of view, a receiver applies meaning to received data in order to produce information.

In every day research contexts, the two words data and information are used interchangeably. Such uses make the assumption that data have meaning; why collect, store, and manage data if they do not have meaning? With having informed the reader of some different perspectives on the definitions of data and information, the remainder of the text defaults to common use of the words and refers to that which we collect, store, and manage in the process of research as data.

Categories of Data

Another perspective through which to understand the essence of data is the many ways in which we categorize different types of data. These categorization schemes each are a model of some aspect(s) of data. For example, Stevens's scales discussed earlier are one such categorization scheme based on the information content of the data. The most helpful categorization schemes are ones that separate the types of data based on some features that are conceptually meaningful or useful in practice. This section covers four such categorization schemes of importance to data collection and processing.

Data have been variously categorized by information theorists and practitioners according to their *topicality*. For a forestry research project, topicality may involve data about trees, data about soil and other environmental conditions, and data about some experimental intervention. At a more generalized and abstract level, some characterize data according to *original observations, derivative data, definitional metadata,* and *operational data*. Original observations are raw data, that is, facts we observe, process, and store about things in the real world. Facts that we calculate or otherwise reason from them are referred to as derivative data. Original and derivative data are often handled differently because given the original data and the rules by which derivative data are generated, derivative data are redundant because they can be recreated. For example, weight and height would be original data, whereas body mass index, here considered derivative data, may be calculated from them. Metadata are often referred to as data about data, for example, a definition or units of measurement. They are often stored differently from original and derivative data and, in fact, the rules by which derivative data are obtained may be considered metadata. Finally, operational data are also considered a type of metadata. These data are generated during the collection, processing, and storage of data. For example, a date and time stamp applied when data are first entered into a system can be considered operational data, as can system logs be generated when rules run to create derivative data or records of changes to data. These categories are informative as we think about collection and processing of data; however, they are not mutually exclusive. For example, date and time stamps are recorded when people log into a data system, whereas operational data may be used in an analysis of system use or in an investigation of system misuse, and thus be considered for these purposes to be original data. Although not mutually exclusive, these categories are useful because methods for collecting, processing, and handling data as well as uses of data differ according to the categories in which they fall for a particular study.

Another perspective (and categorization scheme) based on topicality is categorization of data for a research study as variables versus constants, or as independent, dependent, and covariate variables. The definitions for variables and constants are as the names suggest; *variables* are data elements whose values are expected to change, whereas *constants* are data elements whose values remain the same. Note that this distinction is somewhat artificial because some data elements such as weight may be considered constant in some cases, for example, a short-term study, or dry weight of plant matter at the end of a study, and as changing in other cases, for example, a study of neonates, or a 5-year weight loss study in adults.

Students of research design will recognize categorization of data according to independent, dependent, covariate, and confounding variables. Briefly, a *dependent variable* is the response or outcome variable in a research study. The dependent variable is believed to be influenced by some intervention or exposure. In a study to examine the relationship of stress and hypertension, the dependent variable would be blood pressure. *Independent variables* are the variables believed to influence the outcome measure. The independent variable might be a variable controlled by the researcher such as some experimental intervention or a variable not under the researcher's direct control, like an exposure, or demographic factors such as age or gender. In the example of the hypertension study, the independent variable is stress. *Covariates* are variables other than the dependent and independent variables that do or could affect the response. They can be thought of as variables that if left uncontrolled, they can influence the response but do not interact with any of the other factors being tested at the time. If covariates are present during an experiment, they can affect the study outcome; thus, identifying and controlling (or adjusting for) the covariates increase the accuracy of the analysis. Covariates are observable. *Latent variables,* however, are not observable but may affect the relationship of the independent and dependent variables.

A covariate that is related to both the dependent and independent variables is a *confounder*. This categorization as independent, dependent, and covariate variables is helpful in thinking through study design issues and is best done in the company of a statistician, so that covariates can be identified and controlled for in the design, or measured and adjusted for in the analysis.

Another categorization of data is according to how the data are used. *Primary data* use is the use of data for the purpose for which the data were collected. *Secondary data* use, also called reuse, involves any use of data for purposes other than those for which the data were originally collected. For example, educational data such as student grades collected by school systems and used to generate report cards and determine academic advancement (the primary data uses) may also be used by researchers to evaluate teaching methods or the impact of policy changes in school systems. Here, the research and evaluation are the secondary data use. Another example of secondary data use would be using data from a research project to answer questions not posed in the original study, perhaps years later, and by a different investigator. This categorization is helpful in that it helps make clear that secondary use of data may have some shortcomings and the capability of secondary data to support any particular use needs to be evaluated.

The third categorization of data is called data types. *Data types* are a categorization usually based on the computational operations that can be performed on the data. They have pragmatic significance because they are often leveraged by software to aid in data processing and often are relevant to functionality specific to software that leverage declaration of data types to aid in data handling or processing in some way. As such, many of these operation-based typologies exist. One categorization researchers are likely to encounter (with many variations on the theme) is categorization as *numerical* versus *character* data, that is, numbers versus narrative or labels. Some add date as a separate data type. Categorization according to numerical versus character has some ambiguity because nominal and ordinal data may be stored as numerical values, for example, 1 = yes and 2 = no, where the data have no numerical significance—thus, there is a difference between the data type of the value and the inherent data type. Further, numerical does not distinguish between different types of numerical data such as integers and numbers with decimals.

Some find categorization of data as continuous measures versus discrete measures helpful. A variable that can take on any value between two specified values is called a *continuous variable*. For example, a measured weight of a sample may be 2 ounces, 3 ounces, or any number in between such as a very precise measurement of 2.538 ounces. *Discrete variables* cannot take on any number, for example, the yield from a tomato plant may be one, two, three, and so on tomatoes, but a plant would not grow a half of a tomato. Different statistical operations are appropriate for continuous versus discrete data. A further distinction is made in that discrete variables are sometimes further specified as *dichotomous*, having only two available values, or *polychotomous*, having more than two valid values (Table 2.3).

TABLE 2.3

Some Ways of Categorizing Data

Categorization by topicality: Organization of data elements according to relatedness. For example, a materials researcher may assign a category of *environmental conditions* to all measures of ambient conditions in which his or her experiments are conducted. There are an infinite number of topically based categorization schemes. Topically based categorization schemes in which the categories are conceptually exhaustive over a stable domain of interest and mutually exclusive within that domain will likely be the most useful to researchers.

Categorization by origin: Organization of data elements according to their source on a project. One such common categorization scheme includes original data, derivative data (calculated from original data), definitional metadata (definitional information about original and derived data), and operational metadata (data generated in the collection or processing of original and derived data). A second organization of data elements by origin includes categorization as primary versus secondary data. Primary data are data collected *de novo* for the purpose of the study and secondary data are data collected for some other purpose and reused for a study.

Categorization by stability: Organization of data elements according to the stability of their values over time. For example, variables are data elements whose values are expected to change during a study (or during the execution of a computer program), whereas constants are data elements whose values are stable over the course of a study. True constants such as the speed of light in a vacuum or the atomic weight of an isotope are not expected to change ever.

Categorization by function in a study: Organization of data elements according to the role they play in the planned analysis, for example, dependent versus independent variables, latent variables, confounders, and covariates. A second example of a commonly used categorization includes data elements that are used in the primary planned analysis versus other data. The former are likely subject to a higher level of scrutiny.

Categorization by data type: Organization of data elements according to the fundamental kind of data (Figure 2.2) or by one of the many data type categorization schemes including simple differentiation of character versus numeric data elements, differentiation of discrete versus continuous, differentiation along Stevens's scales, or finer-grained categorization according to ISO 11404 data types or ISO 21090 null flavors.

There are many finer-grained data typologies. The *general purpose data types* put forth by the International Organization for Standardization (ISO) probably are the most representative, in addition to having the authority of being an international standard (ISO/IEC 11404 2007). ISO 11404 defines a data type as a "set of distinct values, characterized by properties of those values, and by operations on those values." As such, it standardizes the name and meaning of common data types. Like the data type categorization schemes described earlier, each ISO 11404 data type is defined based on properties such as computational operations supported, which differentiate the data type from others. As the conceptual definition of data types, ISO 11404 is a reference model for programming languages that use data types. Although ISO 11404 also deals with structural and implementation aspects of data types, we only discuss the former here.

ISO 11404 separates data types into primitives and derivatives. *Primitives* are atomic data types; they are defined explicitly either axiomatically or by enumeration and not composed of or defined based on other data types. Primitive data types defined in ISO 11404 that researchers are likely to encounter in managing research data include the following set. Although many software packages use their own data types, the ISO set is an international standard; thus, some convergence on them is expected. At a minimum, conceptual familiarity with the list of the ISO 11404 data types will help researchers know what to expect.

The *Boolean* is a data type that uses two-valued logic, that is, a dichotomous variable where the two categories are both exhaustive and mutually exclusive. As such the values for a data element of Boolean data type are true and false, where true and false are mutually exclusive. Boolean data types are exact, unordered, and nonnumeric. The operations that can be performed on a data element of Boolean data type include equation, negation, union (or), and intersection (and).

A *state* data type contains a finite set of distinct and unordered values. A data element whose values are considered nominal on Stevens's scales would have a state data type; a list of statuses that a sample might have in a study (collected, processed, frozen, analyzed, and banked) exemplifies the value list for a state data type. State data types are unordered, exact, and nonnumeric, and the only operation state data type support are tests of equivalence (called equation), for example, do two samples have the same state?

Like state data types, *enumerated* data types contain a finite set of distinct values; however, the values in an enumerated data type are ordered. Data elements having an enumerated data type would be considered ordinal on Stevens's scales. Enumerated data types are ordered, exact, nonnumeric, and bounded. Drink sizes—small, medium, and large—are an example of values for a data element of enumerated data type. Note that the value list is bounded, that is, large is the terminal value at one end, whereas small is the terminal value at the other end. Operations supported by enumerated data types include equation, inorder, and successor.

Data elements of *character* data types have values that are members of defined character sets, for example, the English alphabet or Kanji. Character data types differ from state data types in that character data types are dependent on other standards, that is, standardized character sets, whereas state data types are not dependent on other standards for their values. Like state data types, character data types are unordered, exact, and nonnumeric, and the only operation character data type support is equation.

The ISO 11404 *ordinal* data type is the infinite or unbounded enumerated data type. The values of the ordinal data type are first, second, third, and so on to infinity. The ordinal data type is ordered, exact, nonnumeric, and unbounded at both the upper and lower ends. The operations that can be performed on an ordinal data type include equation, inorder, and successor. Data elements having the ISO 11404 character data type could be considered ordinal according to Stevens's scales though the more common conceptualization of ordinal on Stevens's scales is that an upper and lower bounds exist.

ISO 11404 *date-and-time* data types have values representing points in time at various resolutions, for example, years, months, days, hours, minutes, and so on. Values are continuous; thus, two and a half minutes is a valid value, and they are denumerably (countable) infinite. The ISO standard 8601 focuses on representing dates and times using numbers; thus, value spaces for date-and-time data types often conform to ISO 8601. Date-and-time data types are ordered, exact, nonnumeric, and unbounded. The operations that can be performed on data elements of date-and-time data types include equation, inorder, difference, round, and extend. The latter two deal with the precision of a value.

The *integer* data type is the denumerably infinite list of the set of positive whole numbers {1, 2, 3,...}, negative whole numbers {–1, –2, –3,...}, and zero {0}. The integer data type is ordered, exact, numeric, and unbounded. The operations that can be performed on data elements of integer data type include equation, inorder, nonnegative, negation, addition, and multiplication.

The *rational* data type is the mathematical data type whose values are rational numbers, that is, numbers that can be expressed in the form x/y, where x and y are integers and y is not equal to zero. The rational data type is ordered, exact, numeric, and unbounded. The operations that can be performed on data elements of the rational data type include equation, inorder, nonnegative, promote, negation, addition, multiplication, and reciprocal.

Scaled data types have values that are rational numbers resulting from a fixed denominator, that is, each scaled data type is defined by a fixed denominator, and thus, the resulting values are a subset of the rational value space. Scaled data types are ordered, exact, numeric, and unbounded. The operations that can be performed on data elements of scaled data types include equation, inorder, negation, addition, round, multiply, and divide.

The *real* data type is a computational model of mathematical real numbers (number that is either rational or the limit of a sequence of rational numbers). Values of real data types are ordered, approximate, numeric, and unbounded. The operations that can be performed on data elements of real data type include equation, inorder, promote, negation, addition, multiplication, and reciprocal. The precision to which values are distinguishable is important for managing data of the real data type.

Complex data types are computational approximations of mathematical complex numbers. (A complex number is a combination of a real and an imaginary number in the form $a + bi$, where a and b are real numbers and i is the *unit imaginary number* and is defined as the square root of –1.) Values of complex data types are approximate, numeric, and unordered. The operations that can be performed on values of data elements of complex data types include equal, promote, negate, add multiply, reciprocal, and taking the square root. The precision to which values are distinguishable is important for managing data of complex data types.

The ISO 11404 standard allows subsetting and extending standard data types and provides rules for doing so. The standard also permits creation of data types from combining or other operations performed on data types, for example, class, set, bag, sequence, and character string. For these, the reader should refer to the standard directly.

From the descriptions above, it should be evident that the same term found in different sources may not always mean the same thing, for example, ordinal in ISO 11404 and Stevens's scales, and the enumerated ISO 11404 data type is not the same thing as an enumerated data element (Chapter 3). Unfortunately, instances of these are common in the literature pertaining to data management. This is because principles applied in data management come from many different disciplines where the terminology developed independently. Further, with respect to data types, not all software packages support all of the data types described here.

The last categorization of data involves categories of no data. An urban myth illustrates this well. An unfortunate individual jokingly chooses the license plate character string *NO PLATE*. Unbeknownst to him or her, police officers writing out parking tickets write *NO PLATE* in the license plate number field when they come across an illegally parked vehicle with no license plate. The unfortunate individual with the license plate reading *NO PLATE* soon starts receiving many parking tickets. The label *NO PLATE* as used by the police in the story above is an example of a label meant to categorize instances of missing data, in this case signifying on the parking ticket form and subsequently in the data system that the license plate number is missing. There are many categories of *no data*. The most complete enumeration of them is in the ISO 21090 standard. (ISO 2011) The standard defines a *null flavor* as an ancillary piece of data providing additional (often explanatory) information when the primary piece of data to which it is related is missing. The ISO 21090 list of null flavors includes familiar values like unknown, other, asked but unknown, masked, and not applicable among its 14 terms. (ISO 2011) Null flavors are an example of operational metadata. They should be stored in fields other than those reserved for the primary data with which they are associated.

Summary

This chapter presented different perspectives through which we can come to know that which we call data and the stuff that we collect, process, and analyze to answer the scientific questions posed by our studies. The definition of data was explored through semiotic theory as symbols representing something in the real world. Three fundamental kinds of data were introduced and discussed, each in terms of observability and applicable measurement methods. This chapter covered definitions of data and information from multiple perspectives and introduced the concept of data elements as atomic units of information composed of a question–answer format pair and a definition. Additional characteristics of measurable physical quantities such as dimensionality, units of measure, accuracy, and precision were discussed. Common categorizations of data were discussed. Though this chapter was heavy in theory and new concepts, many of them will be used and applied throughout the text.

Exercises

1. A young man gives a young lady whom he is dating a rose on Valentine's Day. (A) Describe the rose in terms of the semiotic triangle. (B) List some data of interest about the rose that a florist might be interested in. (C) On Valentine's Day, there was an outbreak of fever transmitted by rose bugs; describe the data that the public health officials might be interested in. (D) Describe the similarities and differences in the florist and the public health officials model as determined by the data in which they are interested.
2. A young man has a nightmare that his favorite bicycle was stolen. Describe the nightmare in terms of the semiotic triangle.
3. Consider the system of road signs. Describe the system in terms of the four levels of Morris and Stamper's semiotic framework.
4. Consider the system of ground traffic control signals used to direct the movement of planes on the ground at airports. Describe the system in terms of the four levels of Morris and Stamper's semiotic framework.
5. Consider the system of nautical flags. Describe the system in terms of the four levels of Morris and Stamper's semiotic framework.
6. Explain how a date can be stored as a numerical data type and a character data type.
7. A materials science experiment is evaluating four different materials for use in arc furnaces. Arc furnaces use an electrical arc of current moving through air (like lightening) from one surface to another. Each of the four materials is exposed to arcs of different strengths. A device measures the current directly and the material samples are weighed as well as photographed with an electron microscope to assess surface damage. Describe the current, weight, and surface damage observations in terms of Figure 2.3, that is, a prompt, a definition, and a response format. Include valid values if you think any should apply. Associate each data element with the appropriate category on Stevens's scales.
8. An observational research project is studying possible associations of environmental exposure with birth defects. In the study, mothers of infants are asked about different environmental factors to which they are exposed during pregnancy. Describe these observations in terms of Figure 2.3. Associate each with the appropriate category on Stevens's scales.
9. Categorize the inherent information content and the information content on the following form according to Stevens's scales.

6. Date of Birth:______/_____/_______
MM DD YYYY

7. Sex
☐ Male
☐ Female

8. Current Marital Status (please check one):
☐ Married ☐ Divorced ☐ Widowed ☐ Separated
☐ Never Married ☐ Domestic Partner

9. Education Level

Yourself:
☐ Less than High School
☐ High School Graduate (includes equivalent)
☐ Some College or Associate's Degree
☐ Bachelor's Degree
☐ Master's or higher professional degree

10. A forestry researcher is taking a tree survey of an area in which a tree protection experiment is to be conducted. The survey involves measuring and recording the trunk diameter of all trees greater than 6 inches in diameter at a height of 4½ inches from the ground. Explain how the data collected are an incomplete model of reality.
11. Part of the tree protection experiment above requires building temporary gravel roads and walkways through the area. Another member of the research team is in charge of road location. Are the tree measurement data adequate for the second team member's task? If so/if not, why?
12. For each data element on the following form, list (1) the inherent scale (Stevens's scales), (2) the scale of the data collected by the form, and (3) the data type according to ISO 11404.

Primary Infection Site Culture

1 Was a specimen obtained for culture?
☐$_0$ No
☐$_1$ Yes → **If Yes: Date specimen obtained:** ___/___/____
day month year

2 Results of gram stain *(check "negative" or all that apply)*:
☐ Negative ☐ Gram-positive bacilli
☐ Gram-positive cocci ☐ Gram-negative bacilli
☐ Gram-negative cocci ☐ Showing numerous WBCs

3 The result of culture was *(check "no growth" or all that apply)*:
☐ No growth
☐ Contaminant/normal flora
☐ Pathogen isolated → **Specify below.**

4 Pathogen *(complete only if pathogen isolated)*

Type of Specimen	Results (refer to pathogen codes listed in the instructions)	Other (if pathogen not associated with a code; specify Genus and Species)
☐$_1$ Wound drainage ☐$_2$ Needle aspirate ☐$_3$ Tissues ☐$_{98}$ Other, (specify): ________		
☐$_1$ Wound drainage ☐ Needle aspirate		

13. For each data element on the following form, list (1) the inherent scale (Stevens's scales), (2) the scale of the data collected by the form, and (3) the data type according to ISO 11404.

Date of assessment: __ __/__ __ __/__ __ __ __
day month year

Vital Signs

1 Highest temperature: __ __ __ . __ ☐1 °C ☐2 °F } **Check only one:** ☐1 Oral ☐2 Rectal ☐3 Tympanic

2 Blood pressure: __ __ __/__ __ __ mmHg → **Check only one:** ☐1 Sitting ☐2 Supine
systolic diastolic

3 Heart rate: ______ bpm

Assessment of Clinical Signs and Symptoms

Clinical Signs/Symptoms	Assessments		
1 Wound drainage:	☐1 Absent	☐2 Non-purulent	☐3 Purulent
2 Erythema:	☐1 Absent	☐2 Present	
3 Fluctuance:	☐1 Absent	☐2 Present	
4 Localized warmth:	☐1 Absent	☐2 Present	
5 Pain/tenderness:	☐1 Absent	☐2 Present	
6 Edema/induration:	☐1 Absent	☐2 Present	

Primary Site of Infection Evaluation

Record measurements using the largest dimension.

Size of primary infection site:

Length: __ __ __ cm

Width: __ __ __ cm

Clinical Response

Clinical Response *(check only one)*:

☐1 **Cure:** Resolution of clinically significant signs and symptoms associated with the infection present at study admission or improvement to the extent that the infectious process has been controlled and no further therapy with study medication is necessary.

☐2 **Failure:** Inadequate response to study therapy.

☐3 **Indeterminate:** Inability to determine outcome.

14. For each data element on the following form, list (1) the inherent scale (Stevens's scales), (2) the scale of the data collected by the form, and (3) the data type according to ISO 11404.

Sample log sheet for test strips

Date	Cartridge Tested	Test Strip Results mg/L (ppm)	Name of Tester	Notes

15. A researcher categorized data elements according to the following data types: date, text, and numeric. Map these to ISO 11404 data types. If you find that they do not map neatly, explain why.

References

Box GE. and Draper NR. (2007). *Empirical Model-Building and Response Surfaces*, 414.

Checkland P. and Holwell S. (1998). Information, Systems and Information Systems: Making Sense of the Field. Chichester: Wiley.

Davis G. and Olson M. (1985). Management Information Systems: Conceptual Foundations, Structure and Development. New York: McGraw Hill.

Devlin K. (1991). Logic and Information. Cambridge: Cambridge University Press.

Dretske F. (1981). Knowledge and the Flow of Information. Oxford: Blackwell.

Floridi, Luciano, "Semantic Conceptions of Information", The Stanford Encyclopedia of Philosophy (Spring 2017 Edition), Edward N. Zalta (ed.), URL = <https://plato.stanford.edu/archives/spr2017/entries/information-semantic/>.

International Organization for Standardization (ISO) ISO/IEC 11404:2007, Information Technology—General Purpose Datatypes (GPD), 2nd ed., 2007.

International Organization for Standardization (ISO) ISO/DIS 21090, Health Informatics—Harmonized data types for information interchange, 2011.

International Organization for Standardization (ISO). Information technology—Metadata registries (MDR) Part 3—Registry metamodel and basic attributes. ISO/IEC 11179-3:2013.

ISO/IEC. 11179-3:2013 Information Technology Metedata Registries (MDR) Part 3 registry metamodel and basic attributes, 2013.

Merriam-Webster Online Dictionary, entry for data. Accessed on June 16, 2015. Available from http://www.merriam-webster.com/dictionary/data.

Mingers J. (1997). The nature of information and its relationship to meaning. In R. Winder, S. Probert & I. Beeson (Eds.), Philosophical Aspects of Information Systems (pp. 73–84). London: Taylor and Francis.

Mingers J. (2012). Prefiguring Floridi's Theory of Semantic Information. *tripleC* (101). ISSN 1726 - 670X. Accessed May 30, 2015. Available from http://www.triple-c.at.

Morris C. (1938). Foundations of the theory of signs. In O. Neurath (Ed.), International Encyclopedia of Unified Science (Vol. 1). Chicago: University of Chicago Press.

Mosley M. (ed.). *The DAMA Dictionary of Data Management*, 1st ed. Technics Publications LLC, Bradley Beach, NJ, 2008.

Lewis PJ. (1993). Identifying cognitive categories—The basis for interpretative data-analysis within Soft Systems Methodology. *International Journal of Information Management*, 13, 373–386.

Ogden CK and Richards IA. (1923). *The Meaning of Meaning: A Study of the Influence of Language Upon Thought and of the Science of Symbolism*. London: Routledge & Kegan Paul.

Peirce CS. (1931a). A guess at the riddle. In *Collected Papers of Charles Sanders Peirce*, eds. C. Hartshorne and P. Weiss. Vol. 1. Cambridge, MA: Harvard University Press. [Orig. pub. c. 1890.]

Peirce CS. (1931b). Lowell Lecture III. In *Collected papers of Charles Sanders Peirce*, eds. C. Hartshorne and P. Weiss. Vol. 1. Cambridge, MA: Harvard University Press. Lecture delivered 1903.

Peirce CS. (1977). Letter to Lady Welby. In *Semiotic and significs: The correspondence between Charles S. Peirce and Victoria Lady Welby*, ed. C.S. Hardwick and J. Cook. Bloomington, IN: Indiana University Press. Letter written 1908.

Silver G. and Silver M. (1989). *Systems Analysis and Design*. Reading, MA: Addison-Wesley.

Stamper R. (1991). The semiotic framework for information systems research. In H.-E. Nissen, H. Klein, and R. Hirscheim (Eds.), *Information Systems Research: Contemporary Approaches and Emergent Traditions* (pp. 515–528). Amsterdam, the Netherlands: North Holland.

Stevens SS. (1946). On the theory of scales of measurement. *Science* 103(2684): 677–680.

3

Deciding, Defining, and Documenting Data to Be Collected

Introduction

Deciding what data to collect entails decisions about what type of information is needed and the method by which it should be collected. This chapter presents a relevant terminology used in research design, methods of identifying data elements necessary for a research project, and methods for choosing the best data elements in the presence of several options.

Defining data is describing each data element in a way that leaves no ambiguity for humans or computers. This chapter covers the concepts of metadata pertaining to data elements and other attributes necessary to completely specify a data element. The natural tension between limited resources and complete data definition is discussed, as is the difference between data collected for a single use and data that will be reused over time, and by individuals not involved in their initial definition and collection. Methods of documenting and storing data definitions are discussed. This chapter prepares the reader to identify the data elements necessary for a research study, and to choose and apply an appropriate level of data definition and documentation for his or her research.

Topics

- Constructs and concepts
- Conceptual and operational definitions
- Data elements
- Validity and reliability
- Methods for identifying data elements needed for a study
- Collection of additional data

- Precision and significant figures
- Units and dimensionality
- Exhaustiveness and mutual exclusivity
- Defining data elements

Constructs and Concepts

Deciding what data to collect is an early step in the research process. Research usually begins with some question requiring an answer and some idea of how to go about answering it. For example, consider the question, "Do open architectures in public use space result in higher utilization and citizen stewardship?" To answer this question, a design team has obtained two squares at the same intersection with equal access to parking and other relevant amenities. The team constructs an open architecture seating area in one square and a nonopen architecture area in the second. Both spaces are opened to the public for individual use and events. The research question states interest in two things: utilization and stewardship. The research team defines utilization as the number of events for which the space is reserved and the number of person-hours spent in the space. These items are data elements. The research team defines stewardship as an attitude of caretaking and responsibility toward safeguarding another's property. Thus, the team will measure the concepts of event reservations and person-hours, and the construct of stewardship. Identifying and defining concepts and constructs of interest is the first step in deciding what data to collect. Research questions are followed by hypotheses that involve concepts and constructs that are operationalized through data elements (Figure 3.1).

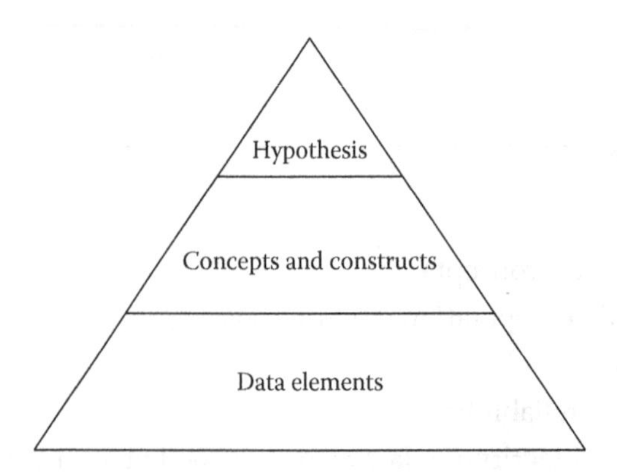

FIGURE 3.1
From hypothesis to data elements.

Both concepts and constructs are abstractions. Human, chair, flower, and dog are abstractions as are ideas of stewardship, beauty, harassment, and type A personality. The difference lies in the underlying phenomena from which the abstractions are made.

Existence of some things is obvious, for example, objects in the real world. A concept is an abstraction formed by generalization from particulars that are known to exist, also called instances (Kerlinger 1999). Consider the concept of an event. Say one of the spaces in the above example was reserved by a concert in the Park organization for a series of eight Friday evening concerts. The particulars are each actual concert and the concept of *concert* is generalized from the key aspects of the particulars that make them members of the *concert* concept and different from other yet related concepts, such as *recital*. The concept *concert* is the universal.

Other things are hypothetical, thought, or implied to exist but cannot be observed or measured either directly or indirectly. Abstractions from these are constructs. For example, a social construct is something that is created by a society, for example, sexism or bipartisanism. Constructs are deliberately created or adopted for a special scientific purpose. For example, a data scientist may rely on the concept of trust to help explain use and nonuse of data from information systems. For example, encountering data errors decreases trust in an information system and those things that erode trust decrease use of such systems. Many constructs are multidimensional. For example, marriage may be composed of loving, partnering, and cohabitating. Although we may be able to count the number of marriages or divorces, measuring marriage itself is different altogether. Constructs, because they cannot be measured directly or indirectly, are measured by asking or by surrogates.

Operational and Conceptual Definitions

Research constructs and concepts need conceptual and operational definitions. A conceptual definition is what you might find in a dictionary; it explains what the concept or construct means, describing the general characteristics and key aspects that distinguish it from other related things. However, in research, operational definitions explain how the concept or construct will be measured. For example, a conceptual definition for water temperature might be something like the following modified from dictionary.com: "a measure of the warmth or coldness [the water] with reference to some standard value" (dictionary.com 2015). The operational definition would read something like the following: "Water temperature is operationalized as the temperature in degrees Celsius as read from the ACME digital Plus model 555-55 thermometer after resting for 2 minutes in a water sample collected by submerging a sampling beaker until full just under the surface of the water and measured while the beaker remains in the water."

Operational definitions should account for situations commonly encountered in practice. An example of an operational definition for tree measurement for tree surveys (Figure 3.2) describes how the measurements should be taken in four commonly encountered situations, a tree on a slope, irregular swelling, multistem trees, and leaning trees. Written instructions may provide even more detail, for example the height above ground at which the measurement should be taken, a diameter below which measurements should not be taken, and direction for handling other scenarios commonly encountered in the field. Operational definitions serve two purposes: First, they help assure that measurements are taken consistently and that data will be comparable, and second, they document how the data were obtained.

For constructs, conceptual and operational definitions work the same. For example, a psychiatric researcher may conceptually define major depressive disorder according to the criteria in the *Diagnostic and Statistical Manual of Mental Disorders* (DSM). The operational definition, however, may read something like this: "current major depressive disorder according to the fifth version of the Diagnostic and Statistical Manual for Mental Disorders (DSM-5) as diagnosed by the Mini International Neuropsychiatric Interview (MINI) and with a score of at least 10 on the Quick Inventory of Depressive Symptomatology—Self-Rated (QIDS-SR; Trivedi et al. 2004) at both screening and baseline visits." Note that as a construct, one of the measurement components of depression is the QIDS-SR questionnaire is completed by patients themselves.

Conceptual and operational definitions are the mechanism by which a researcher increases specificity and traverses the gap between ideas and measures. During and after the study, conceptual and operational

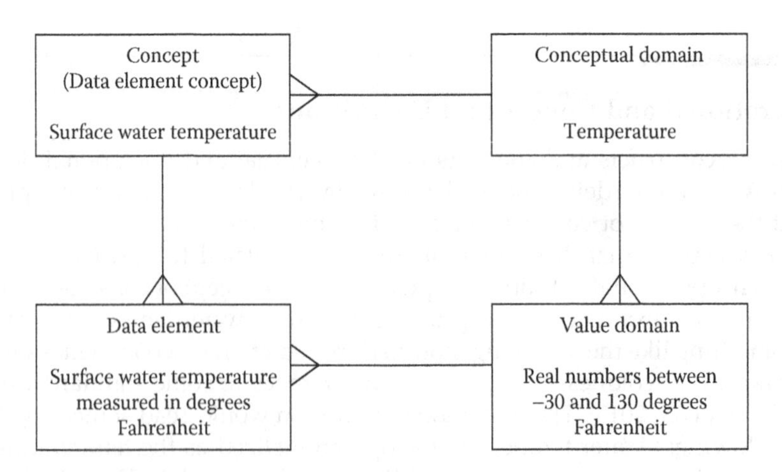

FIGURE 3.2
An ISO 11179 data element.

definitions serve as the Rosetta stone documenting and relate the concepts (or constructs) of interest and the study results. The operational definition should be complete and specific enough so that someone of equal qualifications to the research team but not involved in the research could repeat the measurement. As such, operational definitions are a foundation of research reproducibility.

Repeatability of observations and measurements is important because as we discussed in Chapter 2, operational definitions are a model of the concept or construct that we are measuring. As such, not all operational definitions are equal. Consider the application of the operational definition above for measuring the water temperature of a lake. Submerging the container only until it was full would not provide a valid measure of the water temperature of the body of water because it would only measure the temperature at the surface. If the concept of interest needed to be more representative of the entire body of water, say a depth averaged temperature, use of the surface temperature measurement operational definition would be a poor choice and would lack validity.

Reliability and Validity

Validity and reliability are the properties of data elements or groups of data elements such as in the case of a questionnaire and their operational definitions. *Reliability* is traditionally defined as the degree to which repeated measurements of the same thing return to the same value. Recall that precision is a measure of the extent of reliability. Also recall that accuracy is the extent to which the measured value represents the actual value. A data value can be precise but not accurate and vice versa. Given these, we can say that reliability of a measure (subjective or objective) impacts accuracy, and we might also say that the reliability of the measure sets limits on the achievable accuracy, for example, setting available precision of a numerical measure.

Validity has its origin in measurements of subjective phenomena. *Validity* is traditionally defined as the degree to which a measurement, as specified by an operational definition, corresponds to the phenomena one is trying to assess. The concept of validity is like the combination of aim and choke with shotguns. Aim, of course, describes the location of the average of all of the hits from one shot and choke describes the spread of the pattern from each shot. A measure should be conceptually on target for a concept (aim), and the scope of the measure (choke) should match the scope of the concept, not too broad or too narrow. For example, if a researcher desired a measurement of use of an outdoor park, a daily count of cars in the parking lot would be too narrow in scope because the measure would not account for use by those coming to the park by walking, by public transportation, or by other means.

Similarly, a definition based on the number of people that booked picnic shelters would be off target because it is only one of many uses of the park.

Reliability is germane to all observations and measurements without regard to the subjectivity or objectivity of the measure. For example, a study in which a researcher is coding and characterizing the concepts of love as mentioned in the current best-selling romance novels (more subjective) through the identification of the passages, where love is mentioned, should be reliable as should the assignment of codes to statements in those passages. For this reason, double independent identification and coding are often used, and inter-rater reliability measured and tracked in qualitative research. Likewise, objective measurements such as measuring water temperature are also often subject to repeated measurements in which the average of several measurements is taken or reliability is assessed.

Validity also applies equally to subjective and objective measures. For example, if *love* for the romance novel analysis is defined as romantic love, but instances of familial love or general compassion for others were included in the operational definition and coding, the measure would lack validity due to the operational definition (or its implementation) being broader than the intended concept of romantic love. This was similar for the objective measure of water temperature described earlier, where a surface temperature measure was not a valid measure of the temperature of the body of water.

Data Elements Revisited—ISO 11179

In Chapter 2, we have defined a data element as a question–answer format pair having a definition associated with units of measure and being described as discrete or continuous with valid values. So far in this chapter, we have added to our thinking about data elements in that there are two types of definitions: conceptual and operational. Data elements, at least those for research, should have both a conceptual and an operational definition.

Applying conceptual and operational definitions to data elements should seem like a bit of a leap—after all, we have not said that a data element is a concept, or a construct, or how data elements might relate to concepts and constructs. In fact, a data element is not a concept or a construct. As defined by ISO 11179, a data element is the association of a concept or construct (called a data element concept [DEC]) with a value domain, that is, the possible values that might be collected about the concept. DECs are often but not always more granular than concepts one might describe in a hypothesis statement. In our architecture study example above, the researcher defines utilization as the number of events for which the space is reserved and the number of person-hours spent in the space. Here, the number of events is a data element as is the number of person-hours. The construct utilization

(U) may be calculated (operationally defined) in terms of total person-hours spent in the space per month through the number of events (ne), event duration (ed), event attendance (ea), and the number of nonevent person-hours (neph) spent in the space. In this case, utilization is operationalized through the equation U = ne + ed + neph. The operational definition of calculated or composite data element utilization thus depends on the operational definitions of the component data elements number of events, event duration, event attendance, and number of nonevent person hours. Further, nonevent person-hours may be operationalized as 12-hour counts of the number of people multiplied by the number of hours each spends in the space, where the 12-hour counts occur on 10 randomly selected days equally weighted between week days and weekends. Thus, both the higher level construct and its component data elements have conceptual and operational definitions. It is easy to see how important the operational definitions are to reuse the data for some other purposes such as a food truck owner evaluating how many people he or she would have access to in that particular location on weekends. A common mistake is to provide operational definitions for study endpoints but not for the data elements that comprise them. For example, a clinical study endpoint may be 30-day mortality, that is, whether research subjects remained alive at 30 days after some event. To determine whether an experimental subject met the endpoint, the event date and death date, two data elements, are required for ascertainment of the endpoint. Even though seemingly simple, both data elements should have an operational definition, for example, if death date is not available in medical records, can death date be obtained from an obituary or next of kin, or is a death certificate required? Further, which event date should be used if a subject had multiple events?

The ISO 11179 standard describes data elements in a four-part framework (Figure 3.3). The four parts include a conceptual domain, a data element concept (DEC), a data element, and a value domain. In this framework, the data element is, as we defined in the beginning of this section, a question–answer format pair. For example, if the data element were *United States five-digit zip code*, the value domain would include the complete list of zip codes in the United States at the five-digit level of granularity. Thus, the value domain contains the valid values for data elements. As we see by this example, there can be more than one list of zip codes, for example, the *zip + 4* adds four digits providing a zip code with more granularity. ISO 11179 standard handles this situation of multiple possible *answer formats* through having a data element and a DEC (top left in the diagram). The DEC might be U.S. geographical location, which there are many different ways to specify the latitude and longitude, state, county, city, zip code, zip + 4, street address, and so on. The pairing up of the DEC and one of these value domains creates a data element (bottom left in Figure 3.2). The DEC United States geographical location is a member of the geographical location conceptual domain as are the many possible value domains. The three touch points on the lines from one box to another are called crow's feet. The the crow's feet on the line from a

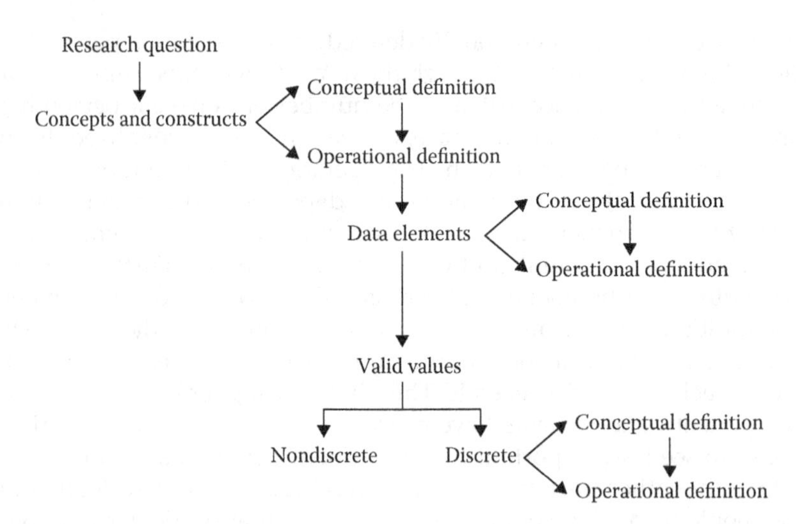

FIGURE 3.3
Applying conceptual and operational definitions to data elements.

conceptual domain to a value domain signify that the conceptual domain may have multiple value domains associated with it, and any particular value domain can be a member of one and only one conceptual domain. Figure 3.3 provides a second example of the data element described earlier for surface water temperature.

Special Considerations for Continuous Data Elements

Let us look at an example with a continuous data element, for example, water temperature. Surface water temperature as defined in the "Operational and Conceptual Definitions" section is the data element concept; it is a member of the temperature conceptual domain. Possible data elements might include the following: (1) surface water temperature measured in degrees Kelvin, (2) surface water temperature measured in degrees Fahrenheit, or (3) surface water temperature categorized as above or below the standard human body temperature. Continuous data elements require specification of the unit of measure. Different units make a different data element because the range of valid values is different where different units are used. Dimensionality is a higher level of abstraction of units of measure. Recall from Chapter 2 that meters per second and miles per hour are different units but share the same dimensionality of length per unit time. Dimensionality is a property at the conceptual domain level.

Continuous data elements also require specification of the level of precision. Precision is determined by the measurement scale used. The final digit recorded is expected to have some uncertainty in it. On a marked scale such as a ruler, the uncertainty is associated with the distance between the two smallest marks, for example, often an eighth of an inch or a millimeter on a household ruler. On such a scale, the last recorded digit should be an estimate of the fraction of the space between marks. Where the distance between the smallest marks is 1 mm, a measured object falling midway between 3.3 and 3.4 cm would be stated as 3.35 where the last significant digit is an estimate. Similarly, on a digital readout, the last digit displayed has some uncertainty. The value on the readout is considered to be ±1 of the last digit, and the last digit on a digital readout is considered significant. Thus, a reading of 3.2 m on a laser measurer means the actual distance is between 3.1 and 3.3 m. This rule applies to numbers measured electronically and stored in a computer directly from the measuring device. However, some software associated with such devices either adds significant digits or is configurable. Either way, the number of significant digits should be determined from the device documentation and applied to measured values. There are rules for handling significant digits with mathematical operations that should also be observed when processing data.

Although continuous data elements do not have a list of discrete valid values, they are often associated with valid values. For continuous data elements, the valid values are stated in terms of the type of number allowed, for example, real, integer, and a valid range, upper or lower limit on some scale determined by the unit of measure.

Specifics for Discrete Data Elements

There are also special considerations for discrete data elements. Valid values (also called permissible values) should be exhaustive over the conceptual space as defined by the conceptual definition of the data element concept. Another way to say this is that the categories must completely cover the concept represented by the data element. For example, if we were collecting data on annual income, we might use the following categories 0–$20,000, $20,001–$40,000, $40,001–$60,000, and so forth up to a large number that encompasses the highest known annual income. This particular example of categories happens to be exhaustive in that it completely covers all possible annual incomes. There is no annual income that cannot be put into one of these categories. If we started at $40,000, however, the list would no longer be exhaustive.

The necessity of having exhaustive categories applies to each discrete data element, for example, both nominal and ordinal data elements. Often for nominal data elements, all of the categories are not needed; in these cases, the list is made exhaustive by the addition of an *other* choice. Although we may not care what the *other* is or even how many *others* exist, the presence of the choice allows the item to be completed by a respondent without forcing one of the other response options that might not fit, or forcing a respondent to leave the item blank, in which case you do not know whether they accidentally skipped without answering or if for whatever reason the item was not applicable. Failing to cover the conceptual space is just a bad data element design.

Valid values for discrete data elements must also be mutually exclusive, that is, the categories should not overlap. For example, if our income ranges consisted of the following categories: 0–$20,000, $20,000–$40,000, $40,000–$60,000, and so forth, respondents making $20,000 and $40,000 per year would not know which response to select. Such overlap means that the values are not mutually exclusive and can render collected data useless.

Likewise, when discretizing a concept into response categories, it is best to mind the level of detail at which each valid value is defined. A consistent level of detail across all of the categories helps convey meaning, and thus helps the respondent make a choice. That is not to say that each valid value must be at the same detail level. For example, if collecting the type of bacteria identified in a culture, for some genera and species might be needed, whereas for other genera may suffice. Note, in this example, the difference in the detail level is deliberate and obvious. Such information reduction may limit data reuse, but does not jeopardize the correctness of the collected data. For any one set of valid values, there may be multiple representations, for example, 1, 2 versus *male, female*. As long as the meaning of the values is the same, regardless of the representation, it is the same value list.

The beginning of this section implied that conceptual and operational definitions are applied to data elements; the ISO 11179 framework explains why and how (Figure 3.2). Conceptual definitions most often describe the data element concept but also apply to members of the conceptual domain and the value domain. Operational definitions add the additional specificity to the data element, including (but not limited to) the valid values and how they are observed or measured. The task, defining data elements for research, necessarily means the latter operational definition and relies on it to be a complete and nonambiguous specification. As such, the prompts, for example, item labels or questions on a data collection form, should be associated with or part of the operational definition. How responses are prompted can influence the response, especially for data that must be obtained by asking.

The ISO 11179 model is a semantic model, the purpose of which is to relate data values to meaning such that these atomic units of information exchange and use, that is, data elements, can be documented, managed, and shared. Humans think in terms of concepts. The ISO 11179 standard provides a model by which symbolic processing performed by computers can occur with the concept as the unit of processing.

ISO and others define the word *concept* in several different ways. Differences and even disagreements over the definition have caused some to abandon the use of the term *concept* completely. Instead, it is better to present and discuss some of the different definitions of the term *concept*. ISO 11179 defines a concept as "unit of knowledge created by a unique combination of characteristics" (ISO 2013). Yet another ISO definition is in the ISO 1087 standard, and it defines the concept as "a unit of thought constituted through abstraction on the basis of characteristics common to a set of objects" (ISO 2000). This latter definition includes mental notions of the term concept such as those at the top of the semiotic triangle (Figure 2.1). Others define the concept as "the embodiment of a particular meaning" (Cimino 1998). Association of both the data element concept and the valid values with the conceptual domain in the ISO 11179 model means that not only is the data element itself associated with a concept, but the valid values for discrete data elements are as well. As such, discrete valid values also must have definitions—conceptual definitions as well as operational definitions. To preserve computability, as described by Cimino, a concept (and thus the symbol or term for it) must correspond to at least one meaning (nonvagueness) and must have no more than one meaning (nonambiguity). Where a meaning corresponds to more than one symbol (redundancy), the synonyms should be associated with the same concept (meaning) (Cimino 1998). Further, within a set of valid values for a discrete data element, we may add that the values need to be exhaustive over the semantic space outlined by the conceptual definition of the data element concept and are mutually exclusive.

When people talk about exchanging data or collecting data, they often speak in terms of fields or variables; usually, they mean the equivalent of the ISO 11179 data element. However, a field connotes a space on a data collection form where a particular data value is to be recorded; the term unfortunately is also used to signify the universal or data element. A variable is a data element whose values change as opposed to a constant (a data element whose values do not change). Change with respect to a variable may pertain to variation from experimental unit to experimental unit, or within experimental units over time. Other than differentiating between constants and values that may change, when the term variable is used, it is synonymous with the term data element—same level of granularity and similar definitional properties are required to uniquely specify a variable.

Deciding What Data to Collect

The concepts discussed thus far in the chapter are critical to systematically deciding what data to collect. Decisions about what data to collect start with the formation of a research question and culminate in data collection instruments, either forms or devices to collect values for specific data elements. Some research takes advantage of multiple types of data collection instruments. Investigators and research teams sometimes use older language and refer to the process of deciding what data to collect as *form design*. However, today when most data are collected electronically directly from devices, data design is a better label. Data design answers the question: *What data are needed to answer the research question(s), how can I completely and unambiguously define them, and what if any relationships exist between them. The latter is covered in Chapter 9?*

The first part of this chapter, although conceptual, provides a systematic method for exploring a research question in terms of the necessary data. First, a research question is stated as a formal hypothesis and is then decomposed into concepts and constructs for which conceptual and operational definitions are written. The operational definitions are further decomposed into data elements for which conceptual and operational definitions are written (or existing data elements already having conceptual and operational definitions meeting the research needs are used). Finally, the discrete data elements are further decomposed into valid values for which conceptual and operational definitions are written if they do not already exist (Figure 3.3).

A similar serial decomposition is used in informatics and related fields to decompose information needs or reports into their component data elements. This serial decomposition method by relying on operational definitions helps researchers identify all of the necessary data, answering such questions as the following: *What level of detail is required? How often and at what time points must the data be collected? What other context is necessary for the correct interpretation of the collected data?*

Serial decomposition can also be applied to cases where existing data are sought to answer a research question. In this case, the research question is decomposed, and the resulting list of operationally defined data elements is compared to existing data sources to evaluate their content. Although there are other considerations as to whether a data set is appropriate such as the population and data quality, content coverage is a necessary first hurdle that must be cleared.

Serial decomposition provides the complete list of data elements necessary for answering a research question. It does not provide any indication of the best representation, format, or structure for collecting the data. These separate concerns are covered in Chapters 5 through 10.

There are other ways to identify data necessary for a study. One such method (a variation on serial decomposition from the research question) is drawing out the tables and figures (including row and column headers, axis labels, subsets, and covariates for which the study might control), which would be used to present the answers to the research questions. These tables and figures can also be decomposed in terms of data elements necessary to populate them. A second method of identifying the necessary data is to look at data element lists, collection forms, or publications from similar studies. Although less systematic, this method may identify confounders, important subsets, or other aspects of the phenomena of interest of which a researcher new to a field might not be aware. A combination of serial decomposition from research questions and an inventory of data that others working in the field collect is usually best for a researcher new to a field. Identifying data to be collected is usually an iterative task that includes review of the list and definitions by others in the field. Such collaborative feedback should culminate in testing of the data collection instruments (forms or electronic devices) to assure that they provide the anticipated information.

Decomposition methods naturally apply Occam's razor to data collection by identifying only the necessary data elements. Asking others and evaluating what others have collected do not. With the latter, the data element list for a study may grow beyond what is feasible. Jacobs and Studer (1991) aptly state that "for every dollar spent to purchase [or create] a form, it requires $20 to $100 to fill it in, process, and store the form." There are, however, very valid reasons to collect more data than strictly necessary for a study. Often studies collect pilot data to aid in the design of the next study, for a related substudy, or for other anticipated secondary uses. Only the investigator can make decisions about the cost versus benefit of collecting additional data.

Deciding what data to collect is best done with the collaboration of a statistician. Both decomposition methods mentioned previously tie data to be collected to the data analysis. Aside from the importance of control and sample size estimation, statisticians are expert in analytic methods, in particular possessing significant knowledge about the assumptions underlying statistical tests and analytical methods. In addition to selecting the best analytic methods for a particular research question, and suggestions for improving the research question, a good statistician can advise in the information content needed to support a particular analysis. A researcher or data manager may do a good job of defining, collecting, and processing their data; however, if they do not have appropriate data to adjust for confounders, or sufficient information content, for example, a scale of *infrequent, occasionally,* and *often*, rather than actual counts, to support the analysis, the data maybe good but somewhat useless. Collaboration with a statistician in the study and data design will help assure useful data.

Choosing the Right Data Element When Options Exist

There are many possible data elements for any concept. There are some special considerations for choosing when multiple possible data elements are under consideration. Some data elements are more readily available than others. For example, if you are conducting research in medical, financial or manufacturing settings, data are likely already collected or being collected. Operationally, it may increase the correctness of the data and the likelihood that data will be collected if consistent data elements are used. In many cases, this means using the data as they are routinely collected by the process being studied.

Consistency in definition, detail level, and representation should be considered. Consistency with other studies in the area may also be a consideration. Secondary use such as pooled analyses or subsets that can be explored when data sets are combined may be important enough to influence the choice of data elements for a study, either use of data elements from a related study or from a relevant standard. The latter, use of standard data elements, controlled terminology, and common data models are increasingly required by some research sponsors. Other considerations that stem from data reuse possibilities include retaining the highest information content, for example, if data exist at a ratio scale maintaining that through data collection and processing, rather than reducing the data to a nominal or ordinal scale will preserve the original information content and support the most reuses of data. And finally but certainly not least is choosing data elements that provide the necessary precision, reliability, and accuracy.

Documenting Data Element Definitions

Documentation of data element definition, aside from a research protocol, may be the most important piece of documentation for a study. Without a clear record of the meaning of the data elements, it is difficult to answer questions about the data after the fact, to reproduce an analysis, or to use data to answer other questions. Others have argued the importance of data definition to research (Muraya et al. 2002, NCES 1999, ICPSR 2012, Strasser et al. 2012). However, there is limited guidance on what constitutes minimum necessary or complete data definition. The application of such guidance is foundational to research replicability and reproducibility and secondary use of data.

The data element is the atomic unit of data collection, use, and exchange as well as the connection between the conceptual and operational realms. Thus, it makes sense that the data element is the prime target for data definition. Database managers, if asked for data definition, will often provide a system, generated data dictionary. Such system-generated lists usually contain a data element label, a data type, and valid values for discrete data elements. Further, such system-generated reports usually describe the data as stored in the system, and thus do not contain information about data elements corresponding to concepts and constructs from the research. Others may provide a data dictionary with information computationally gleaned from the actual data values stored in the system such as the oldest and most recent value for data elements, uniqueness, and descriptive statistics. Neither of these is sufficient because they do not provide full data element definition. During the time after computer data processing was common and up until the past decade, most data were collected on paper forms. It was common practice to write on the form the data element (or variable) name under which the data were stored, the data type, and the stored value or code for discrete data elements. This was called an annotated form. Set in the context of the data collection form with prompts and instructions printed on the form, this practice came close to full data element definition; however, conceptual and operational definition information was often not fully provided by the data collection forms and thus lacking. These annotated forms served as a map between the data collection instrument and the data storage making it easy for those analyzing the data to locate and understand it. Today, when most data are collected directly electronically and where multiple sources of data may exist for a given study, annotated forms are sometimes not sufficient.

For data definition to support data reuse, each data element should have a conceptual and operational definition. If the operational definition does not include a prompt for asked questions, the prompt should be documented. Many studies collect a few hundred or more data elements; thus, applying a unique identifier by which the data elements can be managed and organized is helpful; a name for the data element is often applied as well. The former facilitates tracking, whereas the latter facilitates use of the data elements by humans. Data elements are often categorized variously as described in Chapter 2. At a minimum, data elements should be labeled with a data type. The data type facilitates processing the data values as well as understanding additional and necessary definitional information. The mapping from the data element to the database or other format and structure in which the data are stored or made available for use should be provided. Note that the format in which data are stored during the active phase may differ from that in which data are archived or shared in the inactive phase. Mapping to both formats is necessary, the former to support data use during the study and the latter to support the same afterwards. Discrete data elements should be accompanied by a list of the valid values with conceptual and operational definitions for each valid value. Continuous data elements should be accompanied by their unit of measure, statement of the range or other limits on valid values, and precision. Derived data elements should include in their operational definition the algorithm used for calculation or a pointer to the algorithm. Finally, to assure that when there are changes, for example, data elements added, changed, or no longer collected, the documentation of data definition should be updated and the updates should be tracked. Thus, data definition is not a one-time activity. Such definitional metadata (Table 3.1) is sometimes kept in a spreadsheet, in a data system where data are entered or stored, or in a separate metadata registry.

There is a tension between limited resources and complete data definition. In fact, in a recent survey of academic institutions, lack of funding allocated to data related operations was cited as the primary barrier to implementation of accepted practices in data management. (Keralis 2013) Small or pilot studies may not have the resources to complete and maintain such documentation, grant budgets may not have included funds for data documentation, and institutions often do not provide subsidized or low-cost resources for data documentation, yet it is increasingly being called for. However, because data

TABLE 3.1

Requisite Data Element-Level Definitional Metadata for Research

For all data elements
Conceptual definition
Operational definition
Prompt for asked questions (if not included in the operational definition)
Unique data element identifier
Data element name
Data type
Mapping from the data element to the database or other format and structure in which the data are stored or made available for use
Documentation of any changes to data definition
For discrete data elements
List of the valid values
Conceptual definition for each valid value
Operational definition for each valid value
For continuous data elements
Unit of measure
Statement of the range or other limits on valid values
Precision
For derived data elements
Algorithm used for calculation or a pointer to the algorithm

documentation is foundational to data reuse, and to research replicability and reproducibility, it is hard to justify not defining the data elements for a study and not maintaining such documentation over the course of a study. The only relief to be offered is that for data collected for a single use and not intended for publication or use by individuals other than those involved in their initial definition and collection, less documentation may suffice. However, for data supporting publications, data intended for regulatory decision making, or data intended for reuse, the aforementioned definitional metadata should be considered a requirement.

Summary

In this chapter, we covered concepts critical to the identification and definition of data elements for a research study, a method for identifying necessary data elements, and a method for documenting data element definition.

Resources

1. For more information on concepts, constructs, and their definitions, see text on research design.
2. For more information on reliability and validity, especially how to measure reliability and validity and types of validity, see Edward G. Carmines and Richard A. Zeller, *Reliability and Validity Assessment*, Sage University Paper, 1980.
3. For more information about writing definitions for data elements, see data element definition, Malcolm D. Chisholm and Diane E. Roblin-Lee, *Definitions in Information Management: A Guide to the Fundamental Semantic Metadata*. Design Media Publishing, 2010.

Exercises

1. List the type of variable for the data element *Concomitant medication name.*

   Correct answer: Categorical, polytomous.
2. List the type of variable for the data element *Clinical Gender (male, female, and unknown).*

   Correct answer: Categorical, Polytomous.
3. List the type of variable for the data element *Body Temperature (oral).*

   Correct answer: Continuous.
4. List the type of variable for the data element *Relatedness (not related, probably not related, possibly not related, possibly related, probably related, and definitely related).*

   Correct answer: Categorical, ordinal.
5. An individual does an experiment in the laboratory to identify the best mouse food. They define best as results in highest lean mass, lowest body fat, lowest percent age of myocardial artery stenosis, and lowest cost. In one page or less, describe the following:
   a. Concepts and constructs identified by the research question, including conceptual and operational definitions
   b. Necessary data elements including conceptual definitions, operational definitions, and valid values

6. A research team wants to monitor the migration patterns of wales and see if they follow the migration patterns of their favorite prey. In one page or less, describe the following:
   a. Concepts and constructs identified by the research question, including conceptual and operational definitions
   b. Necessary data elements including conceptual definitions, operational definitions, and valid values
7. A researcher has obtained funding to characterize substance use and abuse in adolescents (type of substance and incidence of use and abuse). In one page or less, describe the following:
   a. Concepts and constructs identified by the research question, including conceptual and operational definitions
   b. Necessary data elements including conceptual definitions, operational definitions, and valid values
8. A researcher is planning an international clinical trial to test the efficacy and safety of chelation therapy in the prevention of secondary cardiac events. In one page or less, describe the following:
   a. Concepts and constructs identified by the research question, including conceptual and operational definitions
   b. Necessary data elements including conceptual definitions, operational definitions, and valid values

References

American Psychiatric Association. *Diagnostic and Statistical Manual of Mental Disorders* 5th ed. Arlington, VA: American Psychiatric Association, 2013.

Cimino J. Desiderata for controlled medical vocabularies in the twenty-first century. *Methods of Information in Medicine* 1998; 37:394–403.

Inter University Consortium for Political and Social Research (ICPSR). *Guide to Social Science Data Preparation and Archiving: Best Practice Throughout the Data Life Cycle* 5th ed. Ann Arbor, MI, 2012.

ISO 1087-1:2000 - Terminology work—Vocabulary—Part 1: Theory ...

ISO 1087-1:2000 Preview. Terminology work—Vocabulary—Part 1: Theory and application.

International Organization for Standardization (ISO). Information technology—Metadata registries (MDR) Part 3—Registry metamodel and basic attributes. ISO/IEC 11179-3:2013.

Jacobs M and Studer L. *Forms Design II: The Course for Paper and Electronic Forms.* Cleveland, OH, Ameritype & Art Inc., 1991.

Keralis Spencer DC, Stark S, Halbert M, Moen WE. Research Data Management in Policy and Practice: The DataRes Project, In Research data management principles, practices, and prospects, Council on library and information resources (CLIR), CLIR Publication No. 160, Washington DC, November 2013.

Kerlinger FN and Lee HB. Foundations of Behavioral Research, 1999. Wadsworth Publishing, Belmont, CA.

Muraya P, Garlick C, and Coe R. Research Data Management. Training materials. Nairobi: ICRAF. 123pp. available from http://www.riccoe.net/linked/muraya_garlick_and_coe_2002.pdf, Accessed April 3, 2017.

Strasser C, Cook R, Michener W, and Budden A. *Primer on Data Management: What You Always Wanted to Know*–2012. Available from http://escholarship.org/uc/item/7tf5q7n3, Accessed June 22, 2015.

Temperature. Dictionary.com. Dictionary.com Unabridged. Random House, Inc. http://dictionary.reference.com/browse/temperature, Accessed June 21, 2015.

Trivedi MH, Rush AJ, Ibrahim HM, Carmody TJ, Biggs MM, Suppes T, Crismon ML et al. (2004). The Inventory of Depressive Symptomatology, Clinician Rating (IDS-C) and Self-Report (IDS-SR), and the Quick Inventory of Depressive Symptomatology, Clinician Rating (QIDS-C) and Self-Report (QIDS-SR) in public sector patients with mood disorders: a psychometric evaluation. *Psychol Med*, 34(1), 73–82.

U.S. Department of Education, National Center for Education Statistics. *Best Practices for Data Collectors and Data Providers: Report of the Working Group on Better Coordination of Postsecondary Education Data Collection and Exchange*, NCES 1999-191, by Melodie Christal, Renee Gernand, Mary Sapp, and Roslyn Korb, for the Council of the National Postsecondary Education Cooperative. Washington, DC, 1999.

4

Data Management Planning

Introduction

Data collection and processing tasks have probably been the most overlooked in research. Historically, these tasks have too often been assumed clerical, not described in research plans, and determined *ad hoc* during the conduct of a study. However, leaving data management planning as an afterthought severely limits options available to investigators and belies allocation of appropriate resources.

Chapter 1 explored the link between reproducibility and data management and introduced the concept of data management plans. A *data management plan* (DMP) is an *a priori* description of how data will be handled both during and after research. A DMP documents data definition and all operations performed on the data, so that the original data can be reproduced from the final data set and vice versa (called *traceability*). The DMP is updated throughout the study as plans change. After the study is over, the DMP provides a record of data definition, collection, processing, and archival. The DMP is to data as an owner's manual is to a car, as such, the DMP is a key component of achieving research replicability and reproducibility, and supporting sharing and secondary use of data.

This chapter explores requirements for DMPs in the United States and other countries, and distills a list of desiderata for DMPs that support research reproducibility, providing the reader a list of contents from which a DMP can be written or evaluated.

Topics

- Defining the DMP
- Example data sharing requirements from around the world

- Contents of a DMP supporting data reuse, research reproducibility and replicability
- Extent of data management planning
- Data flow and workflow.

Data Sharing and Documentation Requirements

DMPs have been variously named and implemented throughout the history of data management. For example, many federally funded studies involving human subjects created a manual of operations or procedures that pertained mainly to data collection in the field. Steps taken toward data analysis were described in a statistical analysis plan. As studies grew in volume and complexity, so did the steps in between the original data collection and the data analysis. When these steps were *just* key entry, many failed to see the need for a plan and record. In fact, most scientific disciplines have only recently articulated the need for or necessary components of DMPs. With data increasingly viewed as potentially supporting research beyond the initial study for which it was collected, research funders and researchers themselves are starting to treat data from studies as an asset of value. This shift in thinking prompts planning and actively managing data resources. Further, as the types and sources of data grow, so does the need for standardization, integration, cleaning, and other processing of data. The number and complexity of steps between data collection and analysis is increasing. Each of these steps performs operations on the data, transforming it in some way, changing values from their original state. Today, reproducing research necessarily entails not just rerunning an experiment but also the data collection or processing, thus, the growing interest in and emphasis on DMPs within the research community. In fact, the Australian Code of Responsible Conduct of Research states, "The central aim is that sufficient materials and data are retained to justify the outcomes of the research and to defend them if they are challenged. The potential value of the material for further research should also be considered, particularly where the research would be difficult or impossible to repeat." And follows on to say, "If the results from research are challenged, all relevant data and materials must be retained until the matter is resolved. Research records that may be relevant to allegations of research misconduct must not be destroyed" (Australian Government 2007).

In 2004, 34 countries including Australia, Canada, China, Japan, the United States, many European countries and others adopted the Organisation for Economic Co-operation and Development (OECD) Declaration on Access to Research Data from Public Funding (OECD 2004). Adoption of the OECD Declaration means agreement with principles (1) that publicly funded research data are a public good, produced in the public interest, and (2) that publicly funded research data should be openly available to the maximum extent possible. These same principles have been in place and even in full implementation, as evidenced by publicly available data repositories, in discipline-based initiatives, for example, Biodiversity Information Facility, Digital Sky Survey, GenBank, International Virtual Observatory Alliance, Protein Data Bank, PubChem, and the Inter-University Consortium for Political and Social Research (ICPSR). These repositories collect data from around the world, and many publishers and research sponsors require submission of data to these discipline-specific repositories, that is, public data sharing.

With sharing of data comes concern for reuse. For example, can secondary users—people not involved in the initial collection or analysis of the data—understand and correctly apply the data to answer new questions? The same concern applies to those seeking to replicate or reproduce a study. Data reuse, research replication, and reproducibility require many aspects of the data to be documented. As a result, many research sponsors have turned to suggesting or all out requiring DMPs.

Today, many large sponsors of research require DMPs of some fashion. The National Science Foundation (NSF) is probably the most notable in the United States. Starting on January 18, 2011, proposals submitted to the NSF required inclusion of a DMP of two pages or less. The NSF DMP is a supplementary document to the project description that according to the NSF should describe how the proposed research will conform to NSF policy on the dissemination and sharing of research results. Although each NSF Directorate has a specific guidance for DMPs, the scope of the NSF DMP as described in the NSF Grant Proposal Guide, Chapter 11.C.2.j *may include* the following (NSF 2011):

- Description of the types of data or other resources resulting from the project
- Data and metadata standards used
- Policies for access and sharing
- Policies and provisions for reuse, redistribution, and production of derivatives
- Plans for archiving and preservation of access

Each NSF Directorate produces guidance specific to its covered disciplines. For example, the Directorate for Computer and Information Science and Engineering (CISE) adds the following things that DMPs in response to their solicitations *should cover as appropriate for the project*:

- Plans for data retention and sharing
- Description of how data are to be managed and maintained
- Detailed analytical and procedural information required to reproduce experimental results (CISE 2015)

Some documents labeled DMPs cover the actual methods by which data were generated, processed, and maintained, whereas others do not (Williams et al. submitted 2016).

Since October 1, 2003, the National Institutes of Health (NIH) has required that any investigator submitting a grant application seeking direct costs of US$500,000 or more in any single year include a plan to address data sharing in the application or state why data sharing is not possible (NIH 2003). Note that the term *data sharing* is significantly narrower in scope than a *DMP*. The NIH data sharing plans cover similar contents as the NSF DMPs with the exception of information about how data were collected and processed. The NIH Office of Extramural Research at NIH suggests that data sharing plans cover the essential information about (NIH OER n.d.)

1. What data will be shared?
2. Who will have access to the data?
3. Where will the data to be shared be located?
4. When will the data be shared?
5. How will researchers locate and access the data?

The examples of data sharing plans provided by NIH are usually less than a page; however, the policy notes that the content and level of details to be included depend on several factors, including whether the investigator is planning to share data, and the size and complexity of the data set.

In the United Kingdom, the seven Research Councils, similar to NSF directorates, also have expectations with respect to data management of those whom they fund. Similar to the NSF, each Research Council in their funding guides or similar information provided for funding applicants or holders specifies these expectations (RCUK 2011). For example, the Science and Technology Facilities Council asks that DMPs "should explain how the data will be managed over the lifetime of the project and, where appropriate, preserved for future re-use." The Biotechnology and Biological Sciences Research Council requires submission of a one-page DMP with the application. Although the Engineering and Physical Sciences Research Council does not require DMPs with research grant applications, their research data principles include that "… project specific data management policies and plans … should exist for all data …"

In Canada, the Canadian Institutes of Health Research requires data deposit for certain types of data, and the Canadian Social Sciences and Humanities Research Council (SSHRC) has a policy requiring that "all research data collected with the use of SSHRC funds must be preserved and made available for use by others within a reasonable period of time," but such sharing is not mandatory, and there are no requirements for data management or sharing plans (SSHRC 2014).

In Australia, having a research data management policy is a requirement under the Australian Code for Responsible Conduct of Research. As of this writing, compliance with the Code, including a DMP, is required at the time of application for the Australian Research Council and the National Health and Medical Research Council funded research, and is under consideration by other funding bodies (ANDS 2011). In compliance with the code, all Australian universities and research institutions have data management policies and guidelines (ANDS 2011). The Australian National Data Service (ANDS) provides guidance to institutions and researchers for complying with the code and describes data management as including all activities associated with data other than the direct use of the data, and goes on to list tasks that data management may include: data organization, backups, long-term preservation, data sharing, ensuring security and confidentiality of data, and data synchronization. ANDS defines data management as including the following activities: data organization, backups, archiving data for long-term preservation, data sharing or publishing, ensuring security of confidential data, and data synchronization. They go on to provide a checklist describing the desired contents of a DMP including the following:

- The data to be created
- Policies applicable to the data
- Data ownership and access
- Data management practices to be used
- Facilities and equipment required
- Responsible party for each activity (ANDS 2011)

Although there seems to be philosophical alignment that research supported from public funds and data collected with public funds should be publically available, just from the few countries represented here, there is significant variability in requirements for data management planning, data documentation, and making research data available to others. Unfortunately, they all conceptualize data management somewhat differently, some concentrating on inactive phase activities including data preservation and sharing, whereas others focus on planning and active phase activities such as data collection and processing. As an aside, the first release of one of the earliest resources for active phase data management planning was produced from the therapeutic development industry in 1994 (ACDM 1996).

There is even greater variability still in what data-related tasks are suggested or required to be included in a DMP (Williams et al. in review 2017). Further, the landscape with respect to data documentation and sharing is changing rapidly and any overview such as this will be out of date by the time it is published. Thus, researchers should always check the requirements for each funder prior to putting together an application and again on award of funding. In the absence of consistency in definition and content of DMPs, this chapter takes a comprehensive approach covering data management from data collection to the point just before the analysis and not including the analysis. Further, this chapter adopts the two goals of data management described in Chapter 1: (1) to assure that data are capable of supporting the analysis, (2) to assure that all operations performed on data are traceable and add a third based on widespread support and even pressure for data sharing, and (3) to facilitate use of data by people other than those involved in their original collection, management, and analysis. This chapter describes creation and maintenance of DMPs to meet these three goals.

What Is in a DMP?

A DMP that supports the three goals of data management is a

1. Plan of what will be done.
2. Store of procedures and guidelines detailing all operations performed on data, including updates to the procedures and guidelines.
3. Record (or guide to the record) of what was done.

As such, the DMP serves three purposes: a planning document, a procedure manual, and a historical record. Building on the construct introduced in Chapter 1, breaking a project into data origination, active phase, and inactive phase (Figure 1.1), data management planning should occur during the planning phase for the study a planning phase. Briefly, the *planning phase* is the period of time from project conception to the start of data collection. The *active* and *inactive phases* are as described in Chapter 1, the period of time for which data accrue and are expected to change, and the period of time following the active phase during which no data accrue or change, respectively. The DMP is started in the planning phase, is maintained through, and serves as the operational resource for the active phase, and serves to document how the data were collected and processed in the inactive phase. It describes plans and procedures for what will be, is being, and has been done. It should be expected (and can be confirmed during an audit) that the procedures and guidelines provided in the DMP comply with applicable regulations and institutional policy. Further, the record of DMP maintenance throughout the project and artifacts from data processing should provide evidence that stated procedures and guidelines were followed.

Different disciplines have evolved specialized practices and conventions optimized for managing data commonly utilized in research in the discipline. Thus, a DMP from one discipline may not be a good fit or template for research in another. At a level of abstraction above these differences, there are topics that any data management plan as defined above should cover (Table 4.1).

TABLE 4.1

Common DMP Topics

- Project personnel, the duration of their association with the project, their roles, responsibilities, data access, training, other qualifications, and identifiers
- Description of data sources
- Description of data flow and workflow from data sources to the final database
- Data element definition
- Data model specifications for active and inactive phases
- Procedures and job aids for operations performed on data, including
  - Observation and measurement procedures
  - Data recording procedures
  - Data processing procedures including description of algorithms
  - Data transfer, receipt, integration, and reconciliation procedures
  - Data quality control procedures
- Description of software and devices used to acquire, process, or store data
  - Calibration plans
  - Configuration specifications for the project
  - Validation or testing plans
  - Scheduled maintenance and plan for handling unscheduled maintenance
  - Change control plans
  - Backup plan and schedule
  - Security plan
- Privacy and confidentiality plan
- Project management plan describing deliverables, completion dates, resource estimates, and description of how progress will be tracked
- Data retention, archival, sharing, and disposal plan

In addition to being driven by scientific needs such as traceability, replicability, and reproducibility, DMP contents are partly driven by ethics, regulatory requirements, and organizational procedures as described in Chapter 1. For example, in the therapeutic development industry, it is common practice to maintain training documentation and other evidence of personnel qualifications such as resumes. The requirement is that an organization be able to demonstrate that personnel are qualified for the tasks they perform on a study. Thus, role-based training and qualification documentation is accompanied by job descriptions connecting qualifications to the roles.

The contents of a DMP are also determined by the types of data managed and the operations performed on them. For example, procedures for importing and checking data from devices are required only if data are acquired from devices for a project. Use of available standards can substantially reduce the amount of documentation required in a DMP. For example, where standard data elements are used, they can be referred to by an identifier with a pointer to the location of the full definition provided by the steward of the standard. In this case, the researcher does not need to describe the complete data element definition, but he or she only needs to point to the standard where it is documented. Similarly, laboratory or institutional procedures that are followed for a particular project can be referenced rather than be redescribed in a DMP, leaving only the description of project-specific procedures requiring full narration in the DMP. Where standards are used, the DMP needs only to serve as the record to a particular version of a standard that is to be, is being, or was used for the project and point to the location where the documentation about the standard can be obtained. Chapter 13 covers organizational versus project-specific procedures. In these cases of standards or institutional procedures being used on a project, the amount of documentation included in the DMP should be based on the extent to which the information is needed by those working on the project. For example, if staff are trained on institutional standard operating procedures (SOPs) and have ready access to them through an institutional document management system, redundantly including them in the DMP provides no additional value. However, if complex data standards are used that, for example, contain many layers of complex and voluminous information, a distillation of the standard may be helpful to staff in understanding data definition, or representation in the information system and in this case should be included.

Examples of plans and procedures included in or referenced by DMPs include data definition, SOPs, data flows, algorithms and code for processing data, work instructions, guidelines, job aids such as checklists, forms, and templates, and other detailed documents supporting SOPs. There are many types of artifacts or evidence of what actually occurred during a project. Examples of these include completed check lists, audit trails of changes to data with change dates and indication of who made the change, approval and signature dates on documentation, system logs, archived files and records, and written reports describing errors that were detected, or instances where procedures could or were not followed and the resulting action taken. The principle is that a complete plan includes a description of each operation to be performed on data in sufficient detail to perform it, and that artifacts produced (and preserved) should evidence the following of the stated procedures.

Throughout a project, any plan or written procedure may need to be updated with a new version to be used going forward, for example, midway through a project, a laboratory instrument reaches the end of useful life as defined by the manufacturer, and a new laboratory instrument is purchased that requires a different type and size of sample container as well as different preprocessing of the sample. Further, in this example, the new containers do not contain a necessary preservative and it has to be added prior to filling the container with sample. In this case, the laboratory SOP or work instruction would be updated mid project to describe the new procedures, approved, and implemented. The old version would of course be retained to document the procedure followed during the date window over which the SOP was valid. This type of change necessarily means that the way of doing some task for the project changed and the updated SOP (or other procedure) documents the change and when it occurred. Continuing with the example, the new instrument required calibration, and the research team needed to know (and show) that the new instrument produced consistent data with the old when the replaced instrument was operating correctly. No SOP existed for installation and qualification of new instrumentation, so the team followed the manufacturer's configuration and calibration procedures, and documented their doing so. To demonstrate that the instrument was calibrated and equivalent to the instrument it replaced, they reran the last 10 samples on the new instrument and reviewed the results side by side with the results from those same 10 samples run on the previous instrument. To document this duplicate sample validation, the team drafted an issue report describing the situation (replacement and qualification of the instrument), how it came about (old instrument reached manufacturer-defined end of life and needed to be replaced in the middle of a study; they needed to be sure that the new instrument was calibrated and working equivalent to the old one), and what was done to resolve it (followed manufacturer-defined calibration procedures and documented results; qualified the new instrument on a series of duplicate samples also ran on the previous instrument). Artifacts of these changes are found in the data, for example, data for the 10 samples run twice the signatures and dates on the SOP; the dates associated with samples run on the new instrument and their new size containers, and the issue report. A summary of how a DMP functions during the planning, active, and inactive data phases of a project is provided in Figure 4.1.

Extent of Data Management Planning

The extent and planning for data management depends on the tasks necessary to collect and process the data and how the data will be used during the project. The number of and relationships between data elements can be few or numerous. The more interrelated the data elements are, the higher the likelihood that at least some of the relationships will need to be enforced and preserved in the data. Such referential integrity is covered in Chapter 9 and checking consistency within a data set is covered in Chapter 10. These aspects of data processing, where required, would be documented in the DMP. As collected, data may be ready to analyze or extensive processing may be necessary. The larger volume of data and more necessary automation and special data processing software are. This is described in Chapters 8, 10, and 12.

At a more basic level, we can also think of data in terms of how, when, and where they move, grow (or shrink), and change over time. These fundamental dynamic aspects of data are covered in Chapter 5. These dynamic aspects of data determine the type of operations that can and should be performed on data, and thus the data processing requirements. The data processing requirements directly translate to procedures to be documented in the DMP.

Start of data collection — Stop of data collection

Planning phase	Active phase (data accumulating and changing)	Inactive phase (no new data, no changes)
Project personnel, the duration of their association with the project, their roles, responsibilities, data access, training, other qualifications and identifiers.	Personnel role and signature log Training and other qualification records	Personnel role and signature log Training and other qualification records
Description of all data sources	Current description of all data sources Record of changes in data sources	Record all data sources
Data and workflow diagrams	Current data and workflow diagrams Record of changes to data and workflow diagrams	Record all data and workflow used
Definition of data elements Planned active-phase data model Planned inactive-phase data model	Current version of data elements Active-phase data model Record of all revisions to data elements	Record of all revisions to data elements Active-phase data model Inactive-phase data model
Procedures for or pointers to procedures for all operations performed on data: Observation and measurement Data recording Data cleaning, transformation and other processing Data integration Data quality control procedures and acceptance criteria	Current procedures (job aid) Record of revisions to procedures Observation and measurement artifacts Data recording artifacts Data cleaning, transformation and other processing artifacts Data integration artifacts Data quality control artifacts Issue reports and resolutions	Record of procedures used Record of revisions to procedures Artifacts from carrying out procedures Issue reports and resolutions
Description of software and devices used for data, including Calibration plans Configuration specifications for the project Validation or testing plans Scheduled maintanance Plan for handling unscheduled maintenance Change control plans Security plan Data backup procedures and schedule	Pointers to system manuals Artifacts demonstrating that procedures were followed	System manuals Record of calibration, validation, maintenance, and change control Summary of back-up and security including incident reports Artifacts from back-up and security activities
Privacy and confidenciality plan	CDAS, DUAs, DTAs, informed consent, or authorization tracking	Association of CDA, DUA, ICF, or authorization to data
Project management plan (deliverables, timeline, tracking reporting plan, and resource estimates)	Record of completion dates	Record of completion dates
The data retention, archival and disposal plan and Data sharing plan items should be in a white stripe/row. They are different than the ones above.	Data retention, archival, and disposal plan Data sharing plan	Data retention, archival, and disposal plan Data sharing plan Curated or archived data and data documentation

CDA: Confidentiality Agreement, DUA: Data Use Agreement, DTA: Data Transfer Agreement, ICF: Informed Consent Form

FIGURE 4.1
Evolution of data management documentation through data management phases.

Data Flow and Workflow

Before written language was developed, early humans used symbols to communicate. Although we are not sure if a picture is really worth a thousand words, we do know that humans can perceive more information and can perceive the information faster from a graphic image than through verbal and written communication channels (Wickens and Hollands 1999). For a research study, it is important to communicate the process through which the data travel from origination to an analysis data set. Graphical depictions are often used to efficiently convey data flow and workflow.

After the data management planning is done, and sometimes as part of the data management planning, the path through which the data travel (called *data flow*), and the tasks performed by humans and machines as part of the data collection and processing (called *workflow*) are diagrammed. Workflow diagrams are also commonly referred to as flowcharts. One-page diagrams, that is, not too detailed, of data flow and workflow show the major components of data collection and processing for a research study. The diagrams illustrate the data collection and processing from two important perspectives: (1) data sources, data movement, and data storage in relation to operations performed on the data (data flow) and (2) sequence and type of operations performed on the data by humans and machines (workflow). Showing both on the same diagram is less helpful because it is easy to confuse the two or show only one without representing the other, for example, showing a data transfer (movement) without showing the steps necessary to affect or complete it. Using two separate diagrams assures that both the data flow and the workflow are completely represented.

There are several sets of standard symbols and conventions in use for creating data and workflow diagrams. The symbols convey meaning; for example, in a workflow diagram, a rectangle is a task and a circle is a continuation symbol used if a diagram spans more than one page. Circles and rectangles have different meanings in data flow diagrams. If symbols are used incorrectly, the diagram will not convey the intended message. Likewise, the conventions can convey meaning; for example, in flowcharts, one direction, that is, top to bottom, implies forward progress of the process, whereas the opposite direction implies looping back or rework. Conventions also help assure that diagrams are well formed and professional. The most popular formalisms for workflow diagrams are Unified Modeling Language activity diagrams and standard flowchart symbols; the latter are the ones found under *shapes* in many software packages. In Chapter 11, this text covers the standard flowchart symbols and related conventions because they are the most common. Data flow and workflow were briefly described here because a list of data sources and required data processing, and data flow and workflow diagrams very much defines the scope of what is covered in a DMP.

Summary

This brief chapter introduced DMPs and explored a variety of definitions or and requirements for DMPs in several countries. A list of contents for DMPs that support traceability and research reproducibility was provided and can be used as a checklist from which a DMP can be written or evaluated. Although this chapter has provided principles and guidance for determining what topics are needed for a DMP supporting reuse, reproducibility and replicability, the actual drafting of the content requires knowledge of the data collection and processing methods, and the ability to match appropriate methodology to a particular data collection scenario. The remainder of this book describes these data collection and data processing basics and fundamental principles that will help researchers select and apply methodology appropriate for a given data collection scenario.

Exercises

1. Which of the following best describes a DMP?
   a. Planned activities
   b. Current descriptions of activities
   c. Revision history of descriptions of activities
   d. All of the above

   Correct answer: (d) All of the above.

2. A laboratory has a standard set of SOPs by which samples are acquired and processed for all projects. Each laboratory member has been trained on the procedures and has access to them through an institutional document management system. Which of the following describes the best way to address the procedures in the DMP for a project?
   a. The procedures should be duplicated and included in the DMP.
   b. The DMP should reference the procedures and state their location.
   c. The procedures do not need to be addressed in the project DMP.
   d. None of the above.

   Correct answer: (b) The DMP should reference the procedures and state their location.

3. Which of the following best describes appropriate handling of changes to a DMP during a project?
   a. There should be no changes to a DMP during a project.
   b. Each team member should update their copy of the DMP to reflect procedures as they perform them.
   c. Changes to procedures referenced in a DMP should be documented throughout the project.
   d. None of the above.

   Correct answer: (c) Changes to procedures referenced in a DMP should be documented throughout the project.
4. Which of the following best describes the relationship of the DMP to regulations?
   a. The DMP should contain a copy of applicable regulations.
   b. The DMP should reference applicable regulations.
   c. Procedures described in the DMP should comply with applicable regulations.
   d. Both (b) and (c)

   Correct answer: (d) Both (b) and (c).
5. A DMP is created during which phase of a project?
   a. Planning phase
   b. Active phase
   c. Inactive phase
   d. None of the above

   Correct answer: (a) Planning phase.
6. A project that involves animal research would NOT include which of the following components of a DMP?
   a. Workflow diagram
   b. Definition of data elements
   c. Privacy and confidentiality plan
   d. None of the above

   Correct answer: (d) None of the above. Even though animals do not have privacy rights, the data from the project may be subject to confidentiality considerations.
7. During which phase of a project should there be no changes to procedures described in a DMP?
   a. Planning phase
   b. Active phase

c. Inactive phase
d. None of the above

Correct answer: (c) Inactive phase. Unless the data sharing, archival or retention plans change.

8. Which of the following DMP components might be changed during the inactive phase of a project?
   a. Data flow diagram
   b. Data quality control procedure
   c. Data sharing procedures
   d. None of the above

   Correct answer: (c) Data sharing procedures.
9. Describe a research project scenario in your discipline. Identify the necessary components of a DMP.
10. For the research project scenario used in the question above, identify the important content for the sections of a DMP.

References

Association for Clinical Data Management. *ACDM Data Handling Protocol Working Party: Data Handling Protocol Guidelines, Version 2.0.* Macclesfield, 1996.

Australian Government, National Health and Medical Research Council, Australian Research Council. Revision of the joint NHMRC/AVCC statement and guidelines on research practice, Australian code for the responsible conduct of research. 2007. Accessed July 1, 2015. Available from http://www.nhmrc.gov.au/index.htm.

Australian National Data Service (ANDS). Data Management Planning 2011. Accessed July 1, 2015. Available from http://ands.org.au/guides/data-management-planning-awareness.html.

Food and Drug Administration. *Guidance for Industry E6 Good Clinical Practice: Consolidated Guidance.* White Oak, Maryland, USA, 1996.

National Science Foundation (NSF). Directorate for Computer and Information Science and Engineering (CISE), Data Management Guidance for CISE Proposals and Awards, March 15, 2015.

National Science Foundation (NSF). NSF Grant Proposal Guide, Chapter 11.C.2.j. NSF 11-1 January 2011. Accessed June 30, 2015. Available from http://www.nsf.gov/pubs/policydocs/pappguide/nsf11001/gpg_index.jsp.

NIH Data Sharing Policy and Implementation Guidance (Updated: March 5, 2003). Available from http://grants.nih.gov/grants/policy/data_sharing/data_sharing_guidance.htm.

NIH Office of Extramural Research (OER). Key Elements to Consider in Preparing a Data Sharing Plan Under NIH Extramural Support. n.d. Accessed June 30, 2015. Available from http://grants.nih.gov/grants/sharing_key_elements_data_sharing_plan.pdf.

OECD. Declaration on Access to Research Data from Public Funding. Science, Technology and Innovation for the 21st Century. *Meeting of the OECD Committee for Scientific and Technological Policy at Ministerial Level*, January 29–30, 2004 - Final Communique. Accessed July 1, 2015. Available from http://www.oecd.org/science/sci-tech/sciencetechnologyandinnovationforthe21stcentury meetingoftheoecdcommitteeforscientificandtechnologicalpolicyatministerial level29-30january2004-finalcommunique.htm.

Research Councils UK. RCUK response to the royal society's study into science as a public enterprise (SAPE), August 2011. Accessed June 30, 2015. Available from http://www.rcuk.ac.uk/RCUK-prod/assets/documents/submissions/SAPE_Aug2011.pdf.

Social Sciences and Humanities Research Council (SSHRC). Research data archiving policy. Date modified: April 8, 2014. Accessed July 1, 2015. Available from http://www.sshrc-crsh.gc.ca/about-au_sujet/policies-politiques/statements-enonces/edata-donnees_electroniques-eng.aspx.

Wickens CD and Hollands JG. *Engineering Psychology and Human Performance*. 3rd ed. Upper Saddle River, NJ: Prentice Hall, 1999.

Williams M, Bagwell J, Zozus M. *Data management plans, the missing perspective*. Submitted to JBI November 2017.

5

Fundamental Dynamic Aspects of Data

Introduction

This chapter opens the discussion on data collection and processing. These are discussed together because decisions about one often require the corresponding adjustments in the other. Historically, investigators and research teams have relied heavily on procedures from previous studies, available tools, or personal experience to design data collection and processing operations for their research. However, in addition to the fundamental types of data (Chapter 2), there are fundamental aspects of how, when, and where data originate, move, change, and grow that should inform decisions about if not determine data collection and processing methods. Further, understanding these relationships in the context of a particular study may influence research design decisions. The principles described here are universally applicable to existing data as well as to data to be collected. These principles should also inform selection of data sources, methods for data collection, and methods for data cleaning.

Two frameworks describing the dynamic aspects of data are introduced to support those who are planning data collection and processing for a research project. Chapters 8 and 10 in Part 2 will concentrate on the types of and the methods for data collection and processing.

Topics

- Time aspects of research design
- Key data-related milestones and their variants
- Mitigating time-related impact on data accuracy
- Data moving, changing, and growing (or shrinking)

Time Aspects of Research Design

Research studies are often categorized as prospective or retrospective. Prospective is essentially looking forward from an event of interest, for example, an exposure or some experimental intervention. Retrospective, then, is looking backward in time from some outcome. In a *prospective study*, the unit of study, for example, a patient, an animal, or a plot of land, has a condition or receives an intervention, and is followed forward over time from the intervention to observe some effect or lack thereof (Agency for Healthcare Quality and Research 2012). Prospective is often taken to imply that data collection starts when the study starts, for example, with the intervention. However, the term applies to the direction of the study, not necessarily the data collection. The parallel and less common terms *prolective* and *retrolective* refer to the timing of data recording. Prolective specifically refers to data recorded after initiation of the study. Retrolective refers to data recorded before initiation of the study. Here, study initiation is usually taken to mean sample selection because the timing of data recording relative to sample selection can be a source of bias (Krauth 2000); for example, selecting a sample after data are collected leaves the potential for knowledge of the data to affect the sample selection. It is unfortunate that the terms prolective and retrolective are not in broader use. Where data will be collected in the future, controls may be added that assure or increase the accuracy of the data. However, recall Figure 1.5, where data have already been collected, and no such opportunity exists; the investigator has no control over how the data are collected and no control over how the data were processed in the past.

Categorization of the research design as prospective or retrospective has sometimes come to be associated with the corresponding intuited impression of data accuracy. For example, data for retrospective and observational studies have often been presumed *dirtier* than data from prospective controlled studies. The problem with such sweeping categorization and reliance on intuition in this case is that prospective studies often include or even rely on data collected in the past, and of course not all observational studies are retrospective or necessarily depend on data that have already been collected. To match research studies with appropriate data cleaning and processing methods, finer distinctions are needed and those distinctions should be based on the characteristics of the data.

Defining Data Accuracy

Accuracy is conceptually defined as the intrinsic property exhibited by a datum (a data value) when it reflects the true state of the world at the stated or implied point of assessment. It follows that an inaccurate or *errant datum* therefore does not reflect the true state of the world at the stated or implied point of assessment. *Data errors* are instances of inaccuracy, any deviation from accuracy no matter what the cause. For example, a problem in a computer program that renders an originally accurate value incorrect is considered to have caused a data error. Data errors are detected by identification of discrepancies through some comparison. A *discrepancy* is an inconsistency, an instance of two or more data values not matching some expectation. Multiple types of comparisons are used in practice to identify data discrepancies. Examples of types of comparisons include comparisons between (1) the data value and a *source of truth,* (2) the data value and a known standard, (3) the data value and a set of valid values, (4) the data value and a redundant measurement, (5) the data value and an independently collected data for the same concept, (6) the data value and an upstream data source, (7) the data value and some validated indicator of possible errors, or (8) the data value and some aggregate statistics. Identification of errors and discrepancies requires some comparison.

Operationally, an instance of inaccuracy or data error is any identified discrepancy not explained by data management documentation (Nahm 2010). The caveat, *not explained by documentation,* is necessary in practice because efforts to identify data discrepancies are undertaken on data at different stages of processing. Such processing sometimes includes transformations on the data (described in Chapters 8 and 10) that may purposefully change data values. In these cases, a data consumer should expect the changes to be documented and traceable through the data processing steps, that is, supported by some documentation.

Key Data-Related Milestones

The relative timing of four milestones in the life of a data value is the major determinant of level of accuracy achievable. The four milestones (Figure 5.1) present virtually in all research projects are as follows: (1) the occurrence of some phenomena or events of interest, (2) the initial recording of data about the event, (3) collection of data about the event for some study, and (4) data cleaning, that is, attempted identification and resolution of data errors (Nahm et al. 2012). Every research study collects data about phenomena or events that occur at some point in time; collectively, these are referred to as occurrence of the event of interest (Figure 5.1). For example, a study correlating prenatal exposure with child development through the first year of life would include observations during the first year of life and information about exposures occurring during pregnancy. The earliest event of interest may have occurred 9 months or more before birth, and the development milestones occur much later. For each event of interest, there is some original recording of data about the event. The original recording may predate and be independent of a research study that may later use the data, or may be part of the study. The original recording of data about childhood development may occur essentially simultaneously with

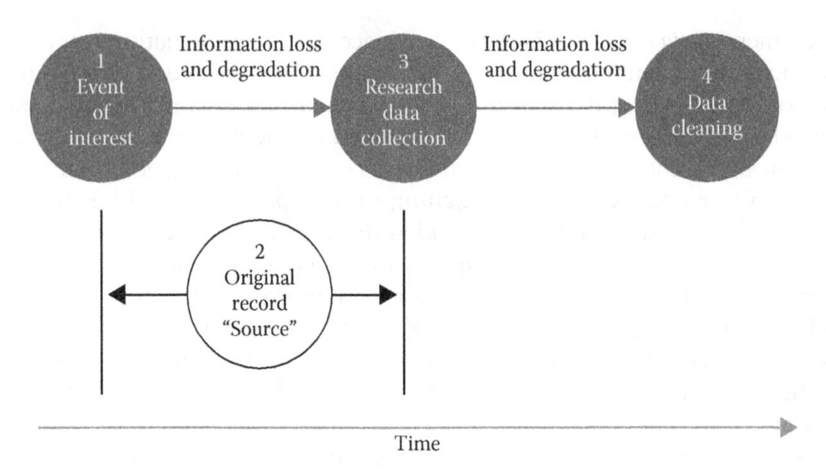

FIGURE 5.1
Key data-related milestones. (1) simultaneous occurrence of the event, data collection about the event and data cleaning, (2) Data cleaning occurs after the event and data collection about the event, (3) both data collection and cleaning occur after occurrence of the event, and (4) Data collection occurs after the event of interest and data cleaning occurs after data collection.

the event; perhaps parents are provided access to an automated phone system where they call in and report the developmental milestones for the study. The prenatal exposure, however, likely was not recorded simultaneously with the exposure. Say, for example, that the study was interested in exposure to household chemicals, and mothers were asked about exposure history after the birth. In this case, the study is *asking* about something that happened up to 9 months before the exposure. Other than the unlikely individual who maintained an inventory of his or her use of household chemicals, the original recording of this information occurs up to 9 months after the exposure occurred, and some information will be lost because recall is likely less than perfect. Availability of a *contemporaneous*, that is, recorded at the same point in time as the occurrence of the event, original recording of data about an event of interest is highly variable and depends on the type of data. Adopting language used in other settings, the original recording of data, whenever it was made, will be referred to as the source. There will not always be a contemporaneous source.

If the events of interest occurred before the start of the study, the researcher has no control over whether or not a contemporaneous source exists, how close to contemporaneous an original recording of the data is, or even if an original recording exists. Such is the case of the study interested in prenatal exposure. However, where the events of interest occur after the start of the study, the researcher has some or even complete control over the data recording and can often arrange for a contemporaneous or near contemporaneous recording of the data.

Regardless of whether a contemporaneous source exists, at some point during or after the event of interest, data will be collected for the study. A gap in time between the occurrence of an event and data collected about an event permits information loss and degradation. If we consider delay in data collection to mean that data are jotted down on some temporary medium such as a scrap of paper or a glove or worse held in memory until they can be recorded, then based on the limitations of human recall, that delay in *data collection* (recording for the purposes of the research) increases the likelihood of inaccuracy. Although there are intervening factors such as the type of event, in general, recall diminishes as the time from the event increases. Data are recorded closer in time to the event are likely to be more accurate.

After collection, data for a research study are often screened to identify potential data errors in the hope of correcting them. Processes to identify and resolve potential data errors are collectively called *data cleaning*. In the ideal case, where data are collected contemporaneously with the event of interest, they can also be cleaned contemporaneously, minimizing information loss and maximizing the likelihood that the error can be corrected. However, data cleaning cannot happen until data are collected, and often occurs afterward.

In any given research scenario, the arrangement in time of these events may take on one of several possible configurations. Most permutations of the three events in time are nonsensical, for example, data collection prior to the occurrence of the event, or identification and resolution of discrepancies prior to data collection. The four arrangements occurring in reality (Figure 5.2) comprise

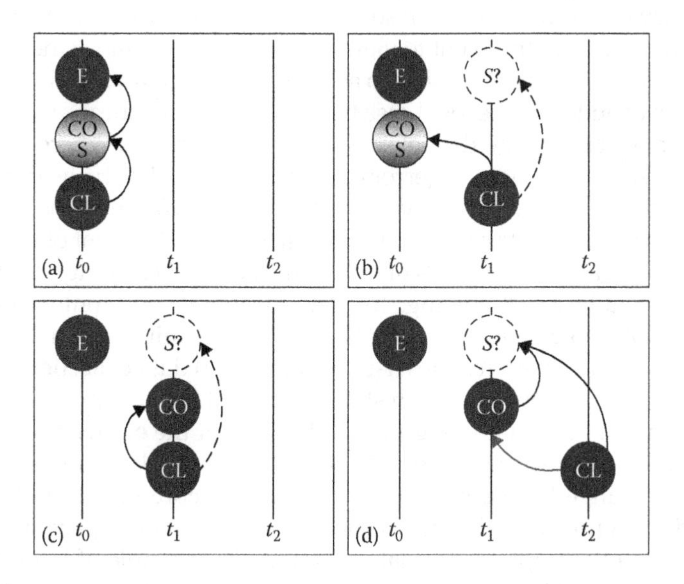

FIGURE 5.2
E: the event of interest, S: the original recording, CO: Collection of the data for a research study, CL: Cleaning of the research data, i.e., identification and attempted resolution of discrepant data. Where t_0 is the time when the event of interest occurs, t_1: some time after t_0, and t_2: some time after t_1. The circles labeled CO S represent a contemporaneous source that can be used in or as data collection. The symbols labeled *S?* represent the possibility of a researcher arranging availability of a source for comparison. The arrows represent comparisons that may be undertaken to identify data discrepancies as part of data cleaning.

distinct cases for data collection and cleaning. These cases are differentiated by the relative timing of data collection and data cleaning with respect to the event of interest and to each other. Usually, more than one of the four cases is possible for any given research scenario, and often, considering these different cases offers different options for data collection and processing, and determines the type of data discrepancies that can be detected and corrected. Therefore, research design decisions that impact or determine key data-related milestones will also determine in part the study's achievable data accuracy.

The Ideal Case

In the ideal case, data collection and cleaning are contemporaneous with the event under study. In this case, the source and the data collection for the study are one and the same. In this ideal case (quadrant *a* in Figure 5.2), all three data milestones occur together in time, giving data discrepancies the maximum likelihood of being identified and resolved while *in* the event. Consider the example of a family photo, when pictures are viewed immediately on a digital camera, they can be retaken when someone's eyes are closed. Waiting for some duration of time after the event to view the picture may remove the possibility of a retake. It is sometimes possible to achieve such contemporaneity in a study. Cleaning is included in the ideal case because even in the ideal case, some loss or errors may occur in the data observation and recording. Where data cleaning is done simultaneously, the errors can be corrected. The truly ideal case, of course, is lossless and completely simultaneous occurrence of the event with data collection - complete capture of the occurrence of the event of interest.

Other temporal arrangements of data collection and cleaning are also common and include the following: (b) data cleaning after contemporaneous event and data collection, (c) data collection and cleaning occurring together but after the event of interest has happened, and (d) data collection occurring later in time than the event of interest, followed still later in time by data cleaning (quadrants b, c, and d in Figure 5.2). Recall that the existence of a source, the original recording, in addition to the data collected for a study may be present or arranged to be present. The source is represented by S in quadrants b–d of Figure 5.2. Existence of such a source further impacts availability and applicability of options for data cleaning. For example, if interviews are recorded, fidelity of the data recorded on the data collection forms during the interview can be measured. If additional accuracy is needed, a redundant and independent data collection process, for example, two recorders, may be pursued. Alternatively, where the interviews are not recorded, or an independent observer is not used, neither measurement nor review will be possible.

In quadrant b, data are collected during the event of interest and are cleaned sometime afterward. An example would be recording responses during a taste test on paper forms, and later entering and cleaning the data. In this case, data entry discrepancies may be corrected by going back to the forms, but errors in recording cannot be corrected and some of them cannot even be detected. There are cases, however, where an independent source can be arranged (*S*? in Figure 5.2). Such an independent source, for example, taking a video or audio recording of the taste test, provides an alternate cleaning path (dotted gray arc in Figure 5.2), and in doing so may increase data quality as if the data were cleaned during the event because the source was preserved.

In quadrant c, data are collected and cleaned simultaneously, but they occur after the event has occurred. Consider the example of interviewing children about what happened on the playground after the recess period has ended and conducting the interview with a computerized form with data error checks running during the interview. In this case, even though the questions asked are about some event in the past, the ability to have real-time discrepancy checks on the computer screen during the interview would prompt the interviewer to ask for clarifications of inconsistencies and provide cues to insure that all needed responses are obtained from the respondent. It may be impossible or prohibitively expensive to recontact the interviewee. Therefore, the capability to provide the interviewer a reminder to reconcile discrepancies during or at the end of the interview is a clear strength. Cleaning simultaneously with collection entails error checks that notify the interviewer of discrepancies or incomplete data during the interview, so that they can be corrected with the respondent. In this case, any discrepancies noted during the interview can be corrected. An independent source might be arranged by taking a video of the children during recess.

In quadrant d, data collection occurs after the event, and cleaning occurs after the collection. This case is the worst scenario. Information loss can occur between the event and the recording, and with cleaning occurring even later, fewer discrepancies can be corrected. Arranging for an independent source in this case usually is at the time of data collection (if the event could have been recorded, doing so would be the better choice, but if data collection must occur after the event, presumably the source must be as well).

From the first principles, the relative timing of data-related milestones impact the data collection and cleaning methods that can be employed in the following ways: In the absence of a contemporaneous source such as an independent recording of the event of interest: (1) collected data cannot be compared with a *source of truth* as we might want to do to identify data discrepancies. If data discrepancies are detected by other means, they cannot be verified. (2) These data cannot be changed with confidence. The problem in both cases is that the source of observed truth (the event itself) has passed and the collected data are all that remain. Without some independent source of the observed truth, for example, an audiovisual or other independent and complete record of the event, the collected data must be accepted as they are, because no basis on which to assess correctness or make changes (corrections) exists.

Any of the arrangements b, c, or d in Figure 5.2 are rendered closer to the ideal where the event of interest is contemporaneously recorded. Such a recording, for example, recording telephone interviews or taking redundant samples, provides an alternative to data collection and cleaning contemporaneous with the occurrence of the event of interest. In the cases of tests run on environmental samples, independent samples can be taken and handled separately, or even sent to completely separate laboratories for analysis, decreasing the risk of data loss due to sample loss, laboratory errors, or other damage. With the necessary resources, collected data or suspected inaccuracies can be verified against the arranged contemporaneous source. The cases above point to actions that can be taken at study planning to prevent or mitigate predictable sources of error.

Time lag between data collection and data cleaning can decrease the achievable data accuracy. For example, prior to the Internet and mobile device-based data collection, some types of data were collected by having the respondent's complete forms that were subsequently sent to a central place for entry. In this scenario, identification and correction of discrepancies was of questionable validity, for example, asking a respondent how well he or she liked the test food 3 weeks after the taste test, and in most cases was not attempted on this basis. Further, such correction after the fact entailed an arduous process of contacting the study sites and the sites in turn contacted the test subjects to ask them about something 3 weeks prior the taste test. Due to recall issues, such cleaning was misguided; thus, respondent-completed forms were not usually *cleaned*. With the Internet and mobile device-based data collection, researchers have the ability to include onscreen discrepancy checks where appropriate and other data cleaning or integrity algorithms such as logging the time of entry. This relocation of the discrepancy detection and resolution process further upstreams in many cases at the point of collection as is common place with mobile technology and gives a better chance of cleaning the data.

The further removed data collection is from the occurrence of the event of interest, and the further data cleaning is from the occurrence of the event of interest and the data collection: (1) the fewer options available for preventing or mitigating, identifying, and resolving discrepancies, and (2) the more resources required to achieve levels of data accuracy obtained in the ideal contemporaneous case.

Data Growing (or Shrinking) Over Time

Similar to how timing of data milestones impacts data cleaning, the aspects of how data change, grow (or shrink), and move also suggest if not determine data management and tracking requirements. Data can grow in multiple ways. Without some training in data management, it can be difficult to prospectively know what types of growth are likely to occur during a study and even more difficult to recognize different types of growth as such when they occur. Similarly, skill is required to match data management methods necessary to appropriately handle the growth.

Consider a simple case where the researcher gets data *as is* that have been collected at some time in the past, for example, spectral data for the 30 seconds following detonation of an explosion. These data are recorded once, and will never be added to and are unlikely to be cleaned or changed in any way. As such, these data require very little in terms of data management other than their definition, initial receipt, quality assessment, documentation, and storage for some period of time. Other research endeavors, for example, an experimental study, collect data as experimental units are accrued over time. Data are collected at multiple time points as longitudinal observations are made. Further, the data collected include multiple data elements, the collection of which may cease or start anew over the duration of the study. For example, in a recent long-term study, one questionnaire was discontinued in favor of another question set a year into the study because the new questionnaire that had just been developed and validated had better psychometric properties. In this example, the number of experimental units and observations on them increases over time and the measures change. These are the fundamental ways that data grow: (1) addition of data values within the existing structure of a study, (2) addition of new experimental or observational units over time, (3) addition of new measures over time, and (4) addition of new time points at which data are collected. The opposite of each are the analogous ways that data can shrink over time, for example, through inactivation of data values, experimental or observational units, measures, or data collection time points. The four ways that data may grow (or shrink) can be thought of as four dimensions (Figure 5.3). Consider a three-dimensional space (a cube). Addition of new values can be represented as the presence or absence of data in the cube signified by the shaded and open circles (Figure 5.3a). Thus, it is a dimension, but not shown on an axis because these changes happen within the existing or planned data structure. The planned data structure is the cube defined by the number of experimental or observational units, the measures on each unit, and the time points over which the measures have been planned to occur. Each of the other fundamental ways in which data grow are shown on the x, y, and z axes representing data collection time points (the x-axis), experimental or observational units (the y-axis), and measures (the z-axis).

Addition or Inactivation of Data Values

Data values are often collected over some period of time. Most data systems facilitate creation of data structures with a *place* for each expected data value. As the data are collected, the data are entered into the data system. Likewise over time, may change (discussed in greater detail in the

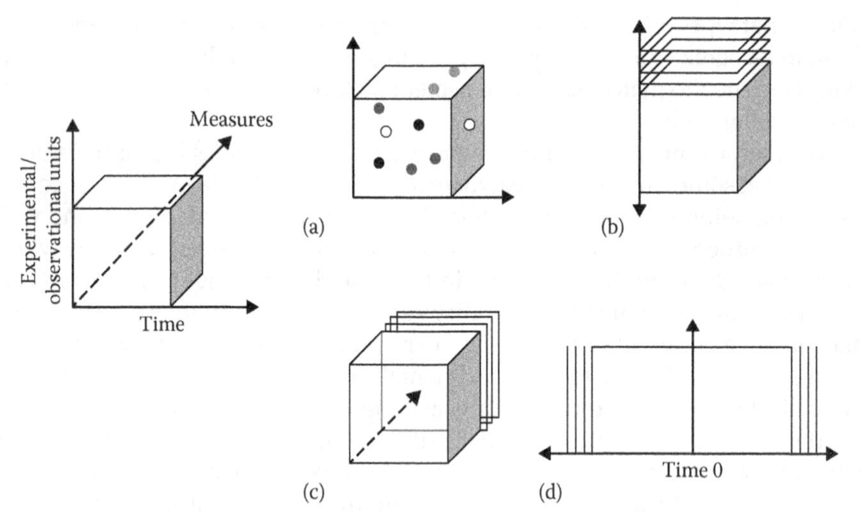

FIGURE 5.3
Ways data grow and change: (a) add or change data values, (b) add (or inactivate) experimental or observational units, (c) add (or inactivate) measures, and (d) add (or inactivate) time points.

chapter section called Data Moving and Changing Over Time) and may be deleted. Most modern data systems employ the concept of a *soft delete,* where the data value is inactivated and would not show on the screen or in reports, but the data system retains a record that the data value once existed. Methods for tracking changes and deletions are discussed later. Figure 5.3a depicts such addition of and changes to data values. These are the fundamental ways in which research data grow (or shrink). Most data systems for research support this type of data growth, although the planning for and tracking of such growth often differs between data systems.

Addition of Experimental or Observational Units

One way that data can grow (or shrink) over time is through the addition of experimental or observational units (Figure 5.3b). Examples include respondents to an online survey that is open for 3 months or the capture and tagging of animals for a telemetry study. In these cases, units are accrued over time as opposed to a situation where all of the data are collected at one point in time on all units and then processed. Here, all of the units are not present on day one; they are added over the days, months, and sometimes years of a study. Planning for accrual of experimental or observational units (and data about them) over time during a study requires decisions about what contextual information is required, for example, enrolment date, situational, and environmental factors at the time of enrollment.

Addition of Measures

Another way that data can grow (or shrink) over time is through the addition of measures or data elements over time during the study (Figure 5.3c). This is the case where a researcher becomes aware of or otherwise needs to add to the data collected, that is, adding the collection of new data elements, or the example above, where an improved questionnaire became available. Sometimes, during a study, the science changes, that is, a new highly relevant measure becomes available and the researcher wants or needs to add it. Sometimes, a measure developer seeks out a study and asks if the new measure can be added in addition to a similarly existing measure for validation purposes. Some long-term studies plan for this to happen and for additional data elements to be collected on new experimental units; however, this situation is often unplanned and is something that an astute data manager will recognize and account for. Sometimes, when a new measure is added, collection of other measures stops. Addition of new measures and discontinuation of existing measures should always be weighed against the impact on the analysis. Most often this occurs in discussions with the study statistician.

Addition of Time Points

The third way that data can grow (or shrink) over time is through the addition or attrition of time points (Figure 5.3d). For example, longitudinal studies make observations over time on the same experimental units. Sometimes, the measures are at regular or prespecified time points, whereas other studies are event driven and data collection is triggered by the occurrence of a specific event. For example, imagine quantifying species density through 50-year-old field succession on 25 plots in different locations around the world. Species counts (observations) may be made on plots (experimental units say randomized to different forestry management strategies) each year. Planning for such data growth over time necessitates tracking time points of the observations, although maintaining the association of observations with the experimental unit on which they were made, and decisions about whether the time points of the observations are planned and whether unplanned observations are allowed. This may also be the case where historical data are gathered, for example, retrospective studies (looking back in time from some outcome of interest) are designed to do this. In some circumstances, a forward looking study may find the need to go after historical data. All of these situations describe planned observations over time. At any point during a study, additional time points may need to be added. For example, in a study, new evidence suggested that measuring the endpoint at 6 months post intervention misses a gradual rebound effect that occurs over the 5 years post intervention. In this case, a study planned to measure study endpoints at 6 and 8 months may choose to skip the eight month time point in favor of yearly measurements through year five.

Data growth can be summarized in three main questions (Table 5.1):

1. Are new experimental units being added over time?
2. Can new measures be added on the existing or new experimental units?
3. How many observation time points are planned?

Where one or more of these growth paths are planned at the time of study design, the data growth they represent needs to be taken into account. Where some type of growth becomes necessary during the study (usually due to external factors such as new information coming available during a study), the astute researcher will recognize the potential impact on management of the data and make necessary adjustments. For example, in clinical studies,

TABLE 5.1

Basic Ways Data Can Grow, Move, and Change

Type of Change	Impact
Are new experimental units be added over time? Up to what point in time? Are there any foreseeable situations that might increase the number?	If new experimental units are expected and the total number may change, data and numbering systems should allow for additions.
How many observation time points are planned for experimental units? Can data be collected at unplanned time points? Are there any foreseeable situations that might necessitate addition of observation time points during the study?	If new time points are expected, data systems should allow for unplanned time points or easy addition of new time points and a mechanism of ordering assessments according to when they have occurred.
Are there any foreseeable situations that might necessitate addition of data elements or measures on the existing or new experimental units?	Data systems should allow for easy addition of new data elements. Data collection may need to be paused while new data elements are added.
Can data values change (or expire) after collection? Under what circumstances can data values change? What type of tracking is required? What type of recovery, that is, restoring data to a certain point in time, is required?	If data changes are expected, data systems should facilitate identification and tracking of changes.
Are data expected to move during the project? Is the movement synchronous (real time) or asynchronous (delayed in time)? Are data moved one case or multiple cases at a time? What volume of data is moved? Are only new or changed data values moved, or is a complete set moved each time? Who maintains the record of changes to data?	If data move, data systems should accommodate the planned movement and integration of incoming data with the existing data.

functionality is often available in data systems to collect study data at unplanned visits, for example, a patient presents with some new problem. When new measures or scheduled observation time points are added during a study, they are added to software used in data collection and management. Table 5.1 provides additional follow-up questions for identifying the type of growth and the potential impact on data collection and management.

Data Moving and Changing Over Time

How data move and change are two additional dynamic aspects of data that impact their collection and management. First, consider the state from which they might move and change. Borrowing a term from the United States Food and Drug Administration (FDA), the original recording of data (or certified copy of the original recording) is called the source. The meaning here is the same as in Figure 5.1. In the Good Laboratory Practices, 21 CFR 58.130 (c) and (e), the FDA requires data to be linked to its source. *Linked* in this case means attributable to the individual who observed and recorded the data, the time of the observation and recording, and that the data should be traceable to the original recording itself. Although the FDA requirements are only applicable to particular studies, the concept of linking data to its source is the foundation of traceability and is very broadly applicable. After the data are first recorded (the original recording), there are reasons that they may change, for example, be corrected, be updated by a current value, or expire. The extent and ways in which data values are expected and allowed to change after their original recording impacts how data are managed. For example, where possible data errors are sought out and attempts made to resolve and ultimately correct them, data systems are often required to track the identified discrepancies, communicate them, make needed updates, and document that a discrepant data value was identified, resolved, and the data value corrected. Planning for such data changes requires decisions about whether changes need to be tracked, and to what extent such tracking should occur, for example, flagging a value as changed or tracking that supports rolling the data back to any point in time. Planning for changing data values also requires explicit rules about the conditions under which data values can change, for example, up to what point in time or by whom. For example, in most studies, there comes a point where all the planned data cleaning has occurred, and data are considered final for the purposes of analysis (or for interim analysis). The more ways in which data are allowed or expected to change over time, the more complex the data management and the more extensive the DMP needs to be.

At the same time, data may move from place to place. Data collected in a researcher's field notes and analyzed on computers in the researcher's laboratory are under the control of the researcher and do not *move*. Alternately, data collected from samples sent to a central laboratory external to the researcher's organization are analyzed in the central laboratory, from which data are subsequently electronically transferred from the laboratory to the researcher *move*. Data originating in an organization may also be sent elsewhere for processing and returned in some enhanced state. Anytime data move across an organizational or system boundary, the complexity of managing the data increases. In the central laboratory scenario above, a contract is likely involved and decisions need to be made about how often and in what format the data are provided to the laboratory and back to the researcher and about which organization (if any) is permitted to make changes to the data, under what conditions, and how those changes are documented. In short, the operations that are to be performed on the data by an external organization need to be completely and explicitly described and are often a matter of contract. The more external data sources, the more complex the data management.

During initial data management planning, it is helpful to show this dynamic aspect of the data in a context diagram (Figure 5.4). A context diagram makes it very clear where data originate, which data crosses organizational boundaries, and where traveling data go. In the example context diagram, materials are tested for a high-temperature use in a plasma trash incinerator. In the researcher's laboratory, each material is subjected to a high-temperature plasma and sent to an external facility where the material surface is examined and surface features are quantified with electron microscopy. Data about material characteristics including composition, production methods, and structural characteristics such as hardness and tensile strength are provided from the manufacturer. The material exposure data and environmental data are obtained and remain within the laboratory

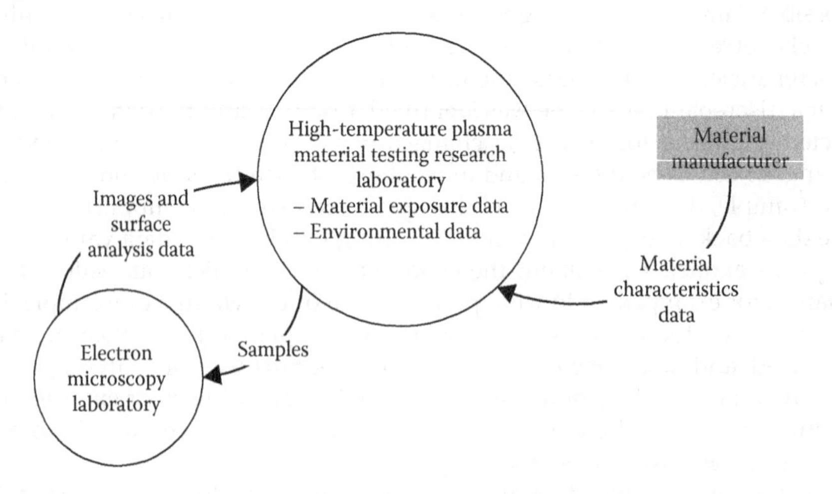

FIGURE 5.4
Example context diagram.

(Figure 5.4). Chapter 11 describes the proper construction of context and data flow diagrams.

Whether and how data move and change during a study in part determine study operational complexity and suggest if not determine the data management methods required. Moving data are data that are collected or processed in one place (or data system) and transferred to another place (or data system). The material testing laboratory example (Figure 5.4) illustrates data in motion. Data in motion can be described as *synchronous* (real time) versus *asynchronous* (delayed in time), by how they are grouped, for example, data for one case versus data for multiple cases together (batch), or by how much data are transferred, for example, a small number of data elements or data values versus a large number of data elements or data values. These differences impact the data system functionality needed and ultimately the extent of data management planning and documentation. In addition, only new or changed data values may be moved (called an *incremental data transfer*) or a complete set may be moved (called a *cumulative data transfer*).

Data moving and changing together suggest if not determine data management methods required. Thus, it is helpful to characterize a study into four quadrants according to how data move and change (Figure 5.5). For example, a simple study using the existing data is often an example of data that are neither moving nor changing. However, a study where a central laboratory is analyzing samples and rerunning tests for damaged samples is an example of data moving (data transferred from the external laboratory to the researcher) as well as changing (reanalysis of damaged samples). In this case, data are both moving and changing. The extent to which traceability is desired will determine the methods and tools used to track the movement of and changes to data.

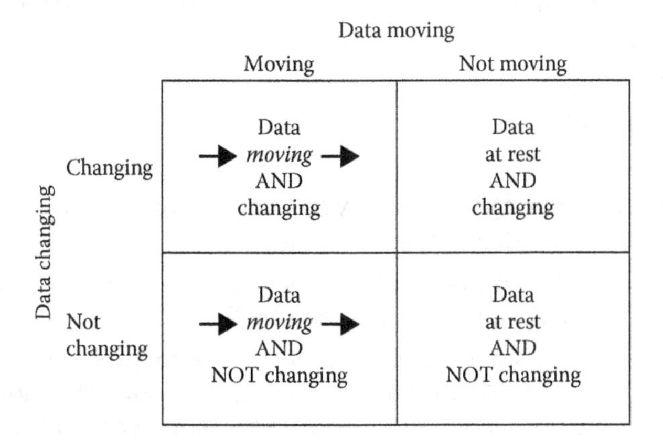

FIGURE 5.5
Combinations of data moving and changing.

Summary

The terms used to describe a study design are not necessarily informative about data accuracy. It is more informative to consider the relative timing of events of interest, the original recording of the events, data collection for the study, and data cleaning (Figures 5.1 and 5.2). Considering these prior to study start and setting up data recoding, collection, and cleaning to be as close as possible to the ideal case (quadrant a in Figure 5.2) will increase the achievable data quality. Data may also be described by how they grow (Figure 5.3), change, and move (Figure 5.5). Where some extent of traceability is desired, these aspects of data determine the complexity of collecting and managing the data. As such, they largely drive the operations performed on data, the methods and tools applied, and the topics that need to be covered in the DMP to adequately document those operations.

Exercises

1. Given the following research scenario, choose the arrangement of key data-related milestones that occurred.

   An educational experiment is being conducted. Two teaching methods are being used in a crossover design in four algebra classes. In four consecutive lessons occurring in consecutive weeks, teaching methods are alternated. The measured outcomes are the weekly test grades. Grades are recorded in the teacher's grade book during grading, and a copy of the tests is sent to the researcher where grades are entered into a spreadsheet and screened for data entry errors.

   a. Quadrant a
   b. Quadrant b
   c. Quadrant c
   d. Quadrant d

   Correct answer: a. Quadrant a. A copy of the test (the source) is preserved and provided to the researcher.

2. Given the following research scenario, choose the arrangement of key data-related milestones that occurred.

   The April census data collection form came to Jan's house by mail in July. The form asks the basic demographic information for people living in the household as of April of the same year. Jan completes the form and mails it back in the provided self-addressed stamped envelope. Data are scanned centrally using character recognition and optical mark recognition. As the form is scanned in the census data center, the scanning software also flags forms with entries that it cannot distinguish with high confidence and flags the form for human verification and subsequent correction.

a. Quadrant a

b. Quadrant b

c. Quadrant c

d. Quadrant d

Correct answer: d. Quadrant d. The event occurred over the prior year ending in April. Data were collected about the event in July. Data are cleaned after the form is received at the data center.

3. Given the following research scenario, choose the arrangement of key data-related milestones that occurred.

A second option exists for census data collection. The form can be completed online by the respondent. In this case, error checks run live, whereas the respondent enters his or her data and the respondent can change or correct responses as he or she enters the data.

a. Quadrant a

b. Quadrant b

c. Quadrant c

d. Quadrant d

Correct answer: c. Quadrant c. Data are still collected after the event occurs, but cleaning occurs during collection.

4. Given the following research scenario, choose the arrangement of key data-related milestones that occurred.

A field test is being conducted of a new surveying instrument. The instrument is deployed to 200 randomly selected surveyors working on road projects in the state. For 2 weeks, the new instrument is used in parallel with the current instrument, that is, measurements are taken with both instruments. Both sets of measurements are documented along with any difficulties encountered with the new or existing technology in the state road projects information system. Data are entered, while the surveyor is still on-site according to standard operating procedures. Basic checks run at the time of entry.

a. Quadrant a

b. Quadrant b

c. Quadrant c

d. Quadrant d

Correct answer: a. Quadrant a. Data are collected during the event (the measurement) or close enough, while the surveyor is still on sight and can remeasure if needed. Data are collected and cleaned also while on sight.

5. Given the research scenario below, choose the optimal timing of data collection and data cleaning.

   A civil engineer is working with the state to measure traffic volume on roads to inform decision making about widening two-lane roads in the state. The engineer has a device that counts cars that his or her supervisor provided. There was no calibration information and no one knew if it had been used before. If the engineer needed to start data collection that day, which of the following choices below would be the best?

   a. Set the instrument up at a road for which traffic volume data are available from similar work 5 years ago.
   b. Forego using the instrument and take manual counts at 15-minute intervals on the hour throughout the day.
   c. Set the instrument up and take manual counts at 15-minute intervals on the hour throughout the day.
   d. Find a second instrument and use both for the first several days.

   Correct answer: c. The manual counts provide a standard by which the instrument can be tested.

6. Explain the advantages and disadvantages of choice (a) above.

   The advantage is that the data offered as a standard are already available. The disadvantage is that five year old data is not a good standard for current traffic volume.

7. Explain the advantages and disadvantages of choice (b) above.

   The advantage in b is the recognition that an instrument that has not been tested should not be used. The disadvantage is that in choice b, the researcher is choosing a manual option in lieu of electronic data collection that with testing could be a more accurate option for the long-term.

8. Explain the advantages and disadvantages of choice (c) above.

   The manual counts provide a standard by which the instrument can be tested. After a period of testing, the more accurate electronic data collection can be used.

9. Explain the advantages and disadvantages of choice (d) above.

   The advantage in d is that an independent instrument/measurement is sought. The disadvantage is that the second instrument likely has not been tested either and as such would not be a good standard.

10. Given the research scenario below, choose the optimal timing of data collection and data cleaning.

    A researcher is trying to assess the impact of different fertilizers on bullfrog populations in the state. He or she has randomized 100 golf courses to three different types of fertilizers. He or she is measuring bull from populations for 3 years. He or she is using frog calls as a surrogate for population levels and is also using local weather monitoring data to capture day and night temperatures in each of the counties. Which of the following is the best choice for data collection?

a. Set timed recording devices at each golf course pond to record sounds from sundown to dawn. Listen to the recordings centrally each month and count bullfrog calls.
b. Recruit 20 data collectors and randomly assign them to attend ponds, count the number of calls per hour, and enter the bullfrog call counts for each hour from sundown to dawn. There would be error checks on the call counts to catch egregious data entry errors.
c. Recruit 20 data collectors and randomly assign them to attend ponds and enter the bullfrog call counts for each hour from sundown to dawn. Have them mark each call as they hear it on a call record sheet. Data would be centrally entered and verified.
d. Set timed recording devices at each golf course pond to record sounds from sundown to dawn. Listen to the recordings daily and count bullfrog calls.

Correct answer: d. The advantage is that a source is preserved ad available to the researcher. The part of d that is an improvement over a is that the recordings and data are attended to daily and any problems, e.g., a malfunctioning recording device, will be identified early.

11. Explain the advantages and disadvantages of choice (a) above.

    Correct answer: The advantage is that a source is preserved ad available to the researcher. The disadvantage in a. is that the recordings and data are only attended to monthly and any problems, e.g., a malfunctioning recording device, will not be identified early; a significant amount of data will be lost if a device malfunctions.

12. Explain the advantages and disadvantages of choice (b) above.

    Correct answer: the disadvantages in b include manual observation, a long retention period (the hour) and manual recording where data cleaning occurs after the hour has passed.

13. Explain the advantages and disadvantages of choice (c) above.

    The advantage of c over b is that the recorder makes a mark for each call so there is no memory retention required. The disadvantage is central data entry and cleaning which occur after the fact.

14. Explain the advantages and disadvantages of choice (d) above.

Correct answer: the advantage in d. is that a source is preserved ad available to the researcher. The part of d that is an improvement over a is that the recordings and data are attended to daily and any problems, e.g., a malfunctioning recording device, will be identified early.

References

Agency for Healthcare Quality and Research (AHRQ). Glossary entry for prospective observational study, accessed May 18, 2012. Available from http://www.effectivehealthcare.ahrq.gov/index.cfm/glossary-of-terms.

Krauth J. *Experimental Design: A Handbook and Dictionary for Medical and Behavioral Sciences* Elsevier, Amsterdam, The Netherlands. 2000.

Nahm M. Data accuracy in medical record abstraction. Doctoral dissertation. University of Texas at Houston School of Biomedical Informatics (formerly School of Health Information Sciences), May 2010.

Nahm M, Bonner J, Reed PL, and Howard K. Determinants of accuracy in the context of clinical study data. *Proceedings of the International Conference on Information Quality (ICIQ)*, November 2012. Available from http://mitiq.mit.edu/ICIQ/2012.

6

Data Observation and Recording

Introduction

Observation, measurement, and recording comprise what many people think of as data collection. This chapter opens with a discussion of the different ways that data values are obtained, including data collected from electronic or mechanical instruments, data collected from existing databases, data abstracted from written documents, data collected by asking people questions, and data collected from observations of entities or phenomena. Differences between manual and automated data acquisition processes are highlighted and used as an organizing construct for the chapter. Methods and special considerations for each are covered. This chapter prepares the reader to plan the data acquisition process for a study.

Topics

- Introduction to manual and automated digital data collection processes
- Observation by humans
- Recording by humans
- Observation and recording by instrumentation

For research studies done from an objectivist perspective, it is important for the observations to be accurate and unbiased. Thus, procedures for making observations and recording data are created to assure that the observations and recordings are accurate, systematic, and consistent between individuals and over time.

Four Possibilities for Observation and Recording

There are two basic options for observation and recording; each can be performed manually or via a device. For example, events such as people passing through a turnstile can be watched and counted by a human or can be quantified using a counter instrument. Similarly, the data—counts of people passing through a turnstile—can be recorded manually, for example, by a human entering the counts into a computer, or can be recorded electronically by the measuring instrument. Together, these form four basic ways in which data are observed and recorded (Figure 6.1). There are different considerations to assure accuracy for manual versus automated tasks such as observation and recording. Thus, it is helpful to consider their implications on data collection and management separately.

Observation by Humans

Human observation is common in research. For example, measuring wildlife populations by counting animal calls or tracks is often accomplished by having a human listen or look for animal sounds or signs. Because of the fact that humans are fallible, are inconsistent, and use judgment in their observations, use of human observation requires standardization and quality control. In the wildlife counting example, calls, tracks, or species may be misidentified causing errors. Observations may be made in fair weather by some observers and in or after rain by others. Inconsistency occurs when the

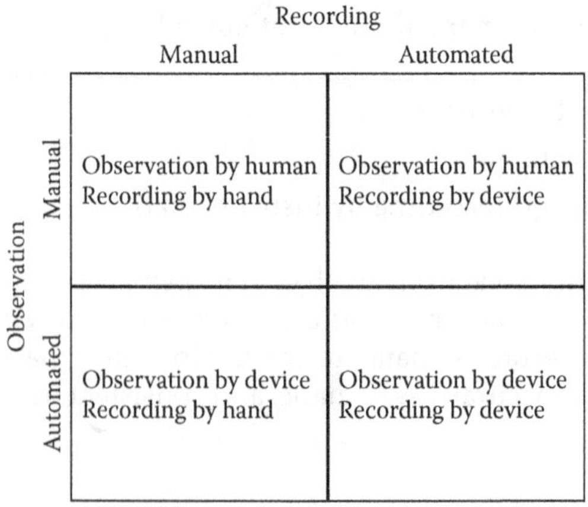

Observation \ Recording	Manual	Automated
Manual	Observation by human Recording by hand	Observation by human Recording by device
Automated	Observation by device Recording by hand	Observation by device Recording by device

FIGURE 6.1
Major variations of data observation and recording.

observers do not use the same rules, for example, how second calls from the same animal or juveniles are counted, or timing and duration of the observations. Standardization of observation procedures is the first step toward increasing consistency in human observation. The goal is that observation be performed the same way by multiple people or by the same person over time. To achieve consistency, all aspects of the observation must be documented and tested to assure that the procedures can be performed consistently by multiple individuals or by the same person over time. If a manual measuring device is to be used in the observation, then the device and its use must also be standardized. Creating such documentation is an iterative process accomplished by writing a first draft, then pilot testing the procedure with several people, and refining the steps to shore up any areas where the testers performed the observation differently. These written standards then become part of the essential documentation describing how the data were collected.

Human observation is complicated further when there is subjectivity or human judgment involved. For example, a sociology student is collecting information about how traffic rules are followed. He or she is counting the number of vehicles that stops at a stop sign and the number that fails to obey the stop sign. Some cars stop at the sign, but others stop further into the intersection, still others slow down but do not stop completely. Different people may make different judgments on these variants, with some people counting stopping 3 feet into the intersection, or slowing significantly as stopping and other people counting these variants as not stopping. Even in this simple case, subjectivity is present. Subjectivity is often present where human observers are asked to classify things or events into discrete categories, to abstract (pick) data out of narrative text, or to observe real-life events. Subjective aspects of observations should be identified, and where the intent is systematic observation, guidelines should be developed for observers on how to *call* each of the subjective cases.

Sometimes, areas of subjectivity and inconsistency are not identifiable in advance. For example, in the stop sign example, on day two of the study, a car stops because a police officer is following it with lights flashing because the driver did not stop at the prior stop sign. The observer was not sure whether that should count as a stop or not. Unexpected areas of subjectivity and inconsistency commonly occur during studies. Thus, standardizing observations in advance is not enough. Observations made by humans require ongoing quality control (Zozus et al. 2015). For human observation, two measures are commonly used: interrater reliability and *intrarater reliability. Interrater reliability* is a measure of agreement between the raters and is calculated when there is more than one observer. Intrarater reliability, however, is a measure of agreement of one rater with himself or herself over time. Percent agreement can be used, or a statistic such as the Kappa statistic that accounts for chance agreement can be used. Assessing inter- or intrarater reliability is necessary whenever there is any subjectivity in observations that are otherwise intended to be objective and systematic, and should occur at multiple points throughout the study (Zozus et al. 2015). There is no one frequency with which inter- or intrarater reliability is assessed throughout a study. As a guide, consider a study in the active phase (ongoing data collection). Four months into the study, an interrater reliability assessment shows that the agreement between the two raters is 62% (pretty low); at this point, the data collected over the prior 4 months would be considered poor and may not support drawing research conclusions. If repeating the prior 4 months of data collection is too big of a loss, then 4 months is too long to wait to assess interrater reliability. As a rule of thumb, the assessments should be at a frequency such that loss of data between two assessments is acceptable. There is also not a specific level below which reliability would be considered too low. Reliability in the upper nineties is good, the higher the better. Publications can probably be found based on reliabilities in the eighties, less so below that. Low reliability can interfere with the analysis because the variability in the observations themselves may be higher than any variability between the compared groups.

Major threats to observations made by humans include human error, inconsistency, and subjectivity. Structure provided by standard procedures, job aids, and training can help prevent these threats but will not eliminate them completely. The extent to which these measures will be effective differs from situation to situation, and cannot be known in advance. Therefore, ongoing assessment of inter- or intrarater reliability is required (Table 6.1).

Obtaining Data by Asking Those Who Directly Observed or Experienced

Obtaining data by direct observation (described above) is different from obtaining data by asking those who directly observed or experienced something. Recall Figure 2.2—some phenomena can be directly observed or measured, whereas others cannot. Obtaining data by asking those who directly observed something is subject to the individual's observation skills, recall, willingness and ability to report accurately, and information loss and degradation that occurs over time. Asking someone who observed something rather than directly observing it for the study often means that the data collection occurs at some later time than the original occurrence of the event of interest and is subject to the information loss and degradation described in Chapter 5.

Asking someone who directly experienced something not directly observable, such as pain or other feelings, is yet again something different. Things that people experience either physically or emotionally are tricky to measure because there is no external standard for comparison and no

TABLE 6.1

Considerations for Data Observed by Humans

Consideration	Implication, Prevention, and Mitigation Strategies
Human error	Humans are fallible. *Prevention strategies include the following*: using a device for the observation if possible. *Mitigation strategies include*: (1) recording the event of interest so that double observation can occur or the recording can be used to assess inter/intrarater reliability for a sample of cases and (2) double independent observation on all or a sample of cases. The weaker mitigation strategy (3) is recording the data real-time where computer error checks can be run during observation and data recording. Strategy (3) is weaker because computerized error checks catch observation and recording errors that cause missing, out of range, and inconsistent but will miss errors that are in range or logically consistent. *Necessary actions*: Inter/intrarater reliability should be measured and reported with research results.
Inconsistency	Human individuality in the way tasks are performed extends itself to data observation and recording. *Prevention strategies include the following*: (1) using a device for the observation if possible, and where not possible, (2) standardizing procedures and testing them so that step-by-step directions are communicated to observers. *Mitigation strategies* are the same as those listed above for mitigation of human error. *Necessary actions*: Inter/intrarater reliability should be measured and reported with research results.
Subjectivity	Many observation tasks involve subjectivity. Inherently subjective phenomena usually require a human observer. *Prevention strategies include the following*: standardizing observation procedures and testing them in advance so that step-by-step directions and guidelines for decision making are communicated to observers. *Mitigation strategies include the following*: (1) recording the event of interest so that double observation can occur or the recording can be used to assess inter/intrarater reliability for a sample of cases and (2) double independent observation on all or a sample of cases. *Necessary actions*: Inter/intrarater reliability should be measured and reported with research results.

way to compare an internal feeling to such a standard. Thus, questionnaires and rating scales are often used. Because of the internal nature of such phenomena (not directly observable), questions asked about them must be validated. Validation is a process that provides some assurance that a question, a question set, or a rating scale actually assesses the desired concept. If a question, a question set, or a rating scale has been validated in some population, the validation information can usually be found in the published literature. Researchers collecting these types of data either (1) use validated instruments, (2) undertake a process to develop them where they do not exist or (3) validate them where they exist, but have not been validated in the population of interest. Data collected using an instrument that has not been validated in a comparable population, and data collection situation may lack the validity and reliability necessary for supporting research conclusions.

For a good discussion of different types of validity, the reader is referred to Carmines and Zeller (1979).

Reliability is often assessed during instrument validation. It is the repeatability of the instrument, that is, if a measure is taken of the same thing in the same situation, the result should be the same (or very close). It measures the closeness of the two (or more) measures under such conditions (Table 6.2).

Internal phenomena that people experience are different from external events that happen in the world, for example, births, deaths, medical procedures, meetings, past exposure to sunlight, or chemicals. Such verifiable facts may be collected and potentially verified or corroborated with external information. For example, if people are asked about deaths of relatives, the information may be verified by researchers using obituary information. Such verification may be planned as part of the study for all or a subset of the population. Where information is subjective or will not be verified, the researcher should strongly consider use of validated instruments. There are often reports in the literature of the accuracy or reliability of information obtained by asking, also called self-reported information or patient-reported information in biomedical contexts.

Recall is also a consideration when asking people about phenomena that they experience. Major life events such as births, household moves, job

TABLE 6.2

Considerations for Data That Are Not Directly Observable and Asked of Others

Consideration	Implication, Prevention, and Mitigation Strategies
Recall	Asking people about events in the past is subject to information loss and degradation associated with the passing of time. *Prevention strategies include the following*: asking questions about the present in the present. *Mitigation strategies include the following*: (1) use of validated instruments, (2) structured interviews such as Timeline Followback that aid recall, and (3) verification or corroboration of facts with external information or independent questions. *Necessary actions*: (1) pointing out limitations of recall-based data in the limitations section of manuscripts and (2) reporting verification results with research results.
Reliability	Reliability is the repeatability of a measure. *Prevention strategies include the following*: using an instrument for which the reliability has been assessed. *Mitigation strategies include the following*: (1) assessing reliability within the study and (2) verification or corroboration of facts with external information. *Necessary actions*: stating what is known about the reliability of questions or instruments in reports of research results.
Validity	Validity is the state achieved when a measure is shown to represent the desired concept. *Prevention strategies include the following*: using an instrument for which the validity has been assessed. *Mitigation strategies include the following*: (1) assessing validity within the study and (2) verification or corroboration of facts with external information. *Necessary actions*: stating what is known about the validity of questions or instruments in reports of research results.

changes, surgeries, and deaths are likely to be recalled with greater accuracy even over long periods of time. For example, in a study that the author was involved with measured accuracy of patient reports of heart attacks, 98% of the reports were affirmatively verified by the reporter's physician. Recall of more mundane events such as dietary intake or vehicle maintenance decreases with time from the event; thus, questions asked to individuals about more mundane events should be asked in the present and about the present rather than about things in the past.

Recording by Humans

Recording by humans requires considerations similar to those for assuring accuracy and consistency in human observation. Like human observation, human recording of data is subject to human error, inconsistency, and recall. Recording of data by humans benefits from prevention and mitigation strategies similar to those applicable for human observation. Further, errors and inconsistency can be prevented by using a device for recording data, for example, a video or audio recording of live events or capture of data electronically from measuring devices. Many measurement devices offer an analog (continuous, such as a dial readout) or a digital display. Although on the surface it may seem easier to have a human record, the data from the display, error, and inconsistency can be prevented by obtaining the data electronically from the device where possible. When reading and recording data values from displays, humans can misread the result or transpose numbers in recording it. Further, humans may inconsistently apply rounding rules or rules for significant figures. All of these can be avoided by obtaining data electronically where possible, by obtaining a preservable print out from the measurement device, or by recording and confirming data contemporaneously (at the time of observation). Such errors can be decreased by standardizing recording procedures, by training data recorders on the standardized procedures, and by assessing adherence to the procedures. Where obtaining data electronically or via a printout from a device is not possible or feasible, human error, inconsistency, and recall issues can be mitigated by (1) making an audio, video, or photo record the event of interest so that double independent data recording can occur or the audio, video, or photo record can be used to assess inter- or intrarater reliability of data recording for a sample of cases, or (2) recording the real-time data where computer error checks can be run during observation and data recording. The latter mitigation strategy is weaker because computerized error checks catch observation and recording errors that cause missing, out of range, and inconsistent but will miss errors that are in range or logically consistent. Where there is a contemporaneous source, for example, a printout from a device or an audio, video, or photo record, recorded data can always be compared with the source to measure a data recording error rate and if needed, corrective action can be taken during the study. The inter- or intrarater reliability and recording error rate should be measured and reported with research results.

There are additional considerations for data recorded by humans centered around cognitive issues of recording data on forms. In seminal work in 1956, George Miller described *The Magical Number Seven, plus or minus two.* Miller (1956) described the capacity of a human to distinguish between a number of things, first distinguishing between tones. He found that with two to five things, the accuracy was quite good; with six or more, the accuracy degraded sharply. This number "seven plus or minus two" is used as the limit of the number of things that humans can remember. This comes into play in recording data in that the number of mental steps or pieces of information that people need to hold in memory while recording data should be as low as possible. For example, having a data recorder sit on the corner and record the number of cars that pass by, making a tick on a sheet of paper or a character in a computer screen is one mental step, that is, making the penstroke or hitting the keystroke. However, data recording tasks can get quite complex, for example, reading a medical record and finding the clotting time laboratory value closest to but not after the procedure time. In the latter case, the data recorder has to hold the procedure time in their memory and screen the medical record for clotting time laboratory values and compare the blood draw time to the procedure time, make a decision, and remember the clotting time value and date or time to record on the form—at least seven mental steps. The mental steps that an individual must undertake is called cognitive load. When forms or screens for data recording and data recording procedures require more than five cognitive steps, errors should be expected. Thus, data recording procedures and forms should be designed such that the raw data are recorded, and human data recorders are not asked to hold values in memory or perform other cognitive tasks such as comparisons, coding, or decision making during data collection. In this latter example, the recording form should collect procedure date or time and all clotting time values and blood draw times on the data form. Computer programs can later be used to make the comparisons and select the correct value. When data observers or recorders are asked to hold values in memory or perform other cognitive tasks such as comparisons, coding, calculating, converting, or other decision making, the recording procedure should be simplified to minimize the number of cognitive steps.

Miller's research also justifies limiting the number of categories to which data recorders must assign values, in that a human data recorder cannot successfully recall the definitions of more than about five categories. Exceptions to this can be rationalized where the data recorders are experts in the subject matter, for example, physicians picking a diagnosis from a hierarchy of 10,000 or more. In this case, the experts group the information in ways that help their recall. For example, the high-level term for heart failure in a common medical terminology set is 150—heart failure, and there are more specific terms including 150.1—left ventricular failure, 150.2—systolic (congestive) heart failure, 150.3—diastolic (congestive) heart failure, 150.4—combined systolic (congestive) and diastolic (congestive) heart failure, 150.9—heart failure, unspecified, and still others where the cause of the heart failure is known. Grouping, also called *chunking* information in this way, helps experts exceed the magic number and handle more information. Thus, where experts are recording data, forms, or screens that leverage common groupings can improve accuracy and speed in data recording. Where forms are designed using groupings in conflict with the expert's common mental models, human performance will be degraded.

From cognitive science, we also know that distribution of information across internal (in the mind) and external (in the world) representations affects human task performance (Zhang 1996). Additionally, representation can extend human performance through external cognition, for example, by limiting the number of things individuals must hold in memory (Zhang and Norman 1994, Zhang 1996). Making information available in the world, such as directions printed on a form, so that respondents or others recording data do not have to remember them, limits the things that data recorders have to remember. Making information available in the world is called distributed or external cognition, and likely improves performance through reduction of the number or extent of mental steps between the data source and the collection form. Things like making definitions or instructions available on the data recording form or screen reduce cognitive load of data recorders and improves accuracy.

The final cognitive science principle with respect to designing forms and screens for human data recorders is the proximity–compatibility principle (Wickens et al. 2016). Although the proximity–compatibility principle pertains to information displays, there are some cases where form or screen design serves both purposes (recording and information display). Adhering to the proximity–compatibility principle means that the form or screen structure should match as closely as possible to the cognitive task that the data recorder must perform. For example, if a researcher is collecting information from tax records and data values from five different tax forms are needed, the data recording form should follow the sequence of the forms in the tax records. This prevents flipping pages or screens back and forth, and allows the recorder to progress forward in parallel through the tax forms and the data recording form without holding values in memory. There may also be cases where a data recorder may be collecting data in real time and have to use the data to make decisions, for example, if a potential research participant is eligible to participate. A simple example of application of the proximity–compatibility principle is phrasing all of the eligibility questions where an affirmative answer indicates eligibility for a particular criteria and a negative answer indicates that the potential participant is not eligible. After the last question is answered, the interviewer needs to just scan down the column of answers, and if all are yes, the subject is eligible. In this example, having different answers indicates that eligibility would require higher cognitive load in determining eligibility and would lead to errors. Of course, this arrangement also makes it easier for a form filler to mark all affirmative answers without reading or thinking about the questions. On electronic forms, downsides such as marking answers can be screened by algorithms that record screen completion time or that prevent forward progress in cases where the form completion time was unreasonably fast.

At the other end of the spectrum from experts, some data recorders are research participants. Like experts, research participants as data recorders deserve special consideration. For one thing, research participants are often everyday people who likely have no training in the subject matter of the research, in data recording, or in use of the specific forms they are completing for a study. Usually, when a research participant fills in a form, it is the first time he or she has seen the form. Thus, more care should be taken to assure that all directions and definitions are printed on the form, that cognitive load on the respondent is as low as possible, and that any cognitive tasks required of the respondent are supported by the form (Table 6.3).

TABLE 6.3

Considerations for Data Recorded by Humans

Consideration	Implication, Prevention, and Mitigation Strategies
Human error	Errors can be *prevented by* (1) obtaining the data electronically from a device where possible and (2) recording data contemporaneously with the observation, and decreased by (3) standardizing recording procedures, (4) training data recorders on the standardized procedures, and (5) assessing adherence to the procedures. Where obtaining data electronically or via a printout from a device is not possible or feasible, human error, inconsistency, and recall issues can be *mitigated by* (1) making an audio, video, or photo recording the event of interest so that double independent data recording can occur or the audio, video, or photo recording can be used to assess inter/intrarater reliability of data recording for a sample of cases and (2) recording the real-time data where computer error checks can be run during observation and data recording. Mitigation option 2 is weaker than option 1. *Necessary actions*: Error rate or inter- or intrarater reliability should be measured and reported with research results.
Inconsistency	Same as above.
Recall	Recording data about events in the past is subject to information loss and degradation associated with the passing of time. *Prevention strategies include the following*: record data when it is observed in the present. *Mitigation strategies include*: Corroboration with independent data. *Necessary actions*: Data should be recorded when the phenomenon is observed. Where contemporaneity and corroboration are not possible, recall should be reported with research results as a limitation to the research.
Cognition	Data collection forms or screens should decrease cognitive load by (1) supplying needed information on the form and (2) supporting cognitive tasks required of the form filler. *Necessary actions*: testing data collection forms.

Lab notebooks and field notes are special cases of data recording. General purpose-bound notebooks are usually blank. No matter what type of data are being collected, there are usually some pieces of information that are expected for each observation. Even cases of nonstructured data such as ethnographic studies often record some structured information; for example, the date, place, and setting are often desirable to record with notes, as are the people involved or their roles. Where the research design calls for structured data, more information such as measurements, units, and notes of any anomalies are recorded for each test article. Where laboratory notebooks or field notes are used for data recording, the required pieces of information should be considered well in advance. It is sometimes helpful to have a stamp made with field labels for the data to be recorded at each observation, or to draw a simple grid to provide some structure. Having a set place for each piece of information can assist the data recorder in remembering each piece of information required, whether just places and dates or a large amount of structured data.

Automated Observation and Recording

Although accuracy, consistency, and systematic methods remain important for data observed and recorded by instrumentation, with instrumentation, there are different threats to data quality and thus different considerations for use of instrumentation in research. Although researchers may not be experts in instrumentation, they must still understand the operable environment and range, the precision, and how the device works so that threats to validity can be recognized and avoided. Researchers must also learn proper operation of the device. Both of these can be obtained from the device manual or manufacturer, as can instructions for how to calibrate and maintain the device, and the necessary calibration and maintenance frequency. The researcher must also be able to confirm that the device is working properly. This is usually accomplished by using the device to measure standards for which the measured values are known in advance, human confirmation of measurements, or comparison to measurements from another device known to be accurate. These are the starting point for using any device in research. Although instrumentation may save time and improve accuracy over manual alternatives, obtaining, learning, assuring consistent use, and confirming a properly working device take time in the beginning of a study and require periodic checks to assure accuracy during the study.

Similar to observation and recording by humans, using instrumentation requires specification of how the instrument is to be configured (values of any user settings) and how the instrument is to be used to make the observation. These procedures should also describe how calibration is assured over time, as well as reference to calibration and maintenance procedures provided by the manufacturer. Members of the research team should be trained in the procedures for use of the instrument in the research. Where instruments are used for observation and measurement, the instrument manual or similar documentation from the manufacturer should be considered an essential document supporting the research. For example, using comparable instrumentation can be important to research reproducibility. Reports of research results should unambiguously state the instrumentation used and any special configuration or settings used fpr the study (Table 6.4).

TABLE 6.4

Considerations for Observation and Recording with Instrumentation

Consideration	Implication, Prevention, and Mitigation Strategies
Malfunction	Devices can come to a researcher broken or can malfunction during a study. *Prevention strategies include the following*: (1) confirming accuracy and reliability of instrumentation in tests prior to use in a study; (2) adhering to manufacturer operating, maintenance, and calibration instructions; and (3) observation strategies that do not rely on only one device. *Mitigation strategies include the following*: graphing data as they are collected, so that unexpected differences or drift are noticed when they occur rather than after the study. *Necessary actions*: prevention strategies 1 and 2, as well as monitoring data as they are collected, and describing these actions with reports of research results.
Improper use	In research settings, it is common to use instrumentation that the researcher does not purchase, has never used before, and for which no manual is provided. This situation invites improper use. *Prevention strategies include the following*: (1) obtaining the manual from the manufacturer, (2) learning how to use the instrument, and (3) testing its use. There are no mitigation strategies for improper use. *Necessary actions*: The manual or operating and maintenance instructions are essential documents for the research and should be archived or otherwise preserved with study documentation.
Representation of precision	Every instrument has a precision. *Prevention strategies include the following*: the precision of the instrument should be appropriate for the research question. The precision of the instrument limits the precision of any subsequent calculations, that is, additional significant figures cannot be added during or after values are used in calculations. There are no *mitigation strategies* for improper representation of precision. *Necessary actions*: The precision of the instrument should be reported with research results.
Accuracy	The manufacturer's stated accuracy of an instrument is available in the manual. Instrument use and lack of calibration or maintenance can adversely impact accuracy. *Prevention strategies include the following*: (1) testing the instrument under the exact same conditions in which it will be used and (2) adhering to manufacturer operating, maintenance, and calibration instructions. *Mitigation strategies include the following*: (1) taking duplicate measurements, i.e., ofn the same sample with a second instrument, and (2) graphing data as they are collected, so that unexpected differences or drift is noticed when they occur rather than after the study. *Necessary actions*: prevention strategies 1 and 2, as well as monitoring data as they are collected, and describing these actions with reports of research results. Where prevention strategies can't be used, the researcher can employ mitigation strategies.

(*Continued*)

TABLE 6.4 (*Continued*)

Considerations for Observation and Recording with Instrumentation

Consideration	Implication, Prevention, and Mitigation Strategies
Sampling rate	Sampling at a rate that is too low results in information loss. *Prevention strategies include the following*: setting the sampling rate such that no useful information is lost. There are no mitigation strategies. *Necessary actions*: using an appropriate sampling rate.
Noise	Noise can come from many sources and can obscure a measured signal. *Prevention strategies include the following*: understanding what the signal of interest should look like and testing instrumentation in the environment in which it will be used. Sources of noise can be identified and eliminated or signal processing strategies such as filtering may be sought. There are no mitigation strategies except after the fact processing of the signal to separate as much as possible the signal from the noise. *Necessary actions*: eliminating noise that may obscure the signal of interest.
Artifacts	Artifacts appear in a measure signal as a result of things that occur during measurement. The author separates artifacts from other noise because artifacts often come from preventable or identifiable causes such as movement during measurement or actions such as powering up the instrument. Once identified, many of these can be prevented or eliminated. *Prevention strategies include*: same as above for noise. There are no mitigation strategies. *Necessary actions*: identifying all artifacts and eliminating sources of artifacts that may obscure the signal of interest.
Chain of custody lapses	Documented chain of custody is necessary to demonstrate that measures were in place to prevent unwanted or undocumented data changes. *Prevention strategies include the following*: setting up a secure chain of custody and documenting all individuals and machines with access to data. There are no mitigation strategies. *Necessary actions*: documenting the chain of custody of all research data.
Lack of traceability	Traceability with respect to research data is the ability to show all changes to data between the original data and the data analyzed for a study. Lack of traceability means that undocumented changes were made to data and bring the integrity of the research into question. *Prevention strategies include the following*: setting up a secure chain of custody and assuring that all changes to data by man or machine are tracked. There are no mitigation strategies. *Necessary actions*: documenting the chain of custody and tracking all changes to data.
Loss of context	Research studies have differing needs for association of data values with context such as time, location, and experimental unit. *Prevention strategies include the following*: identifying and providing for all necessary associations. There are no mitigation strategies; once associations have been lost, the data sometimes cannot be reassociated. *Necessary actions*: identifying and providing for all necessary associations.

Where instrumentation is used to measure one or more parameters at a single point in time, assuring accuracy of the data is fairly straightforward. Proper use of a calibrated and maintained instrument capable of the necessary precision, within the environment for which it was designed to operate and within its operable and range, is the starting point. The wise researcher may also periodically take measurements of standards or samples where the parameter values are known to assess drift over time. Double measurement of split samples, measurement of independent repeat samples, or measurement of split or independent repeat samples on a comparable instrument may help assure that any problems are detected. For example, the measurement of split or repeated samples can detect an instrument that is sensitive to sample placement. In all three cases, the measurements would be expected to match. These are also examples of reliability assessments. Although such assessments may increase time and cost, they are necessary to demonstrate confidence in the measurements.

Although all of these considerations apply to measuring dynamic phenomena over time (or space), some additional considerations apply. Consider a researcher interested in measuring outdoor air temperature. Outdoor temperature outdoors is a continuous function; it changes throughout the day, often, but not always reaching a high each afternoon and a low overnight. If only one temperature reading is captured per day, the low and high and a lot of values in between would be missed. Intuitively, enough measurements have to be taken so that the features of interest are not missed. Say the researcher is interested in temperature changes prior to storms where the temperature may fluctuate between 10°C and 20°C in an hour, and the researcher is familiar enough with the phenomena to know that fluctuations never happen faster than 1°C every 5 minutes. As the researcher is familiar with the timescale on which the phenomena of interest change, he or she can safely decide how often measurements are needed; he or she has the knowledge to weigh the benefits and costs of a more detailed signal against the resolution that may be forfeit with a lower *sampling rate*. Picking a sampling rate requires familiarity with the timescale of the phenomena of interest. The Nyquist–Shannon sampling theorem states a minimum sampling rate (e.g., measurements per second) that will enable faithful representation of a signal with no information loss (Shannon 1949). Application of the theorem requires knowing the highest frequency in the waveform; these are phenomena that happen faster than other parts of the waveform. In the temperature example, if the sampling rate had been set by the theorem assuming a 24-hour cycle, the sampling rate would have been too low to capture the detail in the storm-related transients. The take-away points are that (1) sampling a continuous function at a rate that is too infrequent and results in information loss, and (2) setting a sampling rate requires knowledge of the timescale of the event of interest.

Testing instrumentation in the environment in which it will be used is necessary. Instrumentation is subject to noise, for example, instrumentation in a lab may pick up noise from fluorescent lights or the microwave in the break room. Signals from instrumentation should be examined for noise, and efforts undertaken to reduce the noise without also reducing the data. Measurements taken with instrumentation may also have various artifacts. For example, a researcher may notice a spike at the beginning of the file when the device is initially turned on. Such artifacts may skew statistics calculated from the data.

Familiarity with what the signal should look like is helpful because then artifacts and noise can be recognized, and the researcher can take steps to identify and reduce, or otherwise account for the sources of noise and artifacts.

Many instruments have a mechanism of recording the data electronically. Some require interfacing the instrument with a larger system, whereas other instruments provide more direct access such as writing data files to a portable drive or personal computer. Unfortunately, instruments are not usually built with the knowledge of a research project and how the project names and tracks data. Thus, the researcher usually needs to create a file naming convention, file storage, and a back up mechanism. Often there is not a way to enter a sample number, site number, subject number, or time point into the device to label the data and formally associate the data with important context information. For example, a study for which the author managed data received image files and the identifier for the experimental unit was in the file name but nowhere in the file itself. This is often the case when using instrumentation in research. The context to be associated with each file (and data within the file if necessary) should be identified and documented so that important associations can be maintained. See Chapter 9 for discussion of data structures and maintaining such associations.

Similarly, because instruments are sometimes selected for a research study and not part of established infrastructure, the researcher must establish chain of custody for the data received from the device. Chain of custody is a concept from law enforcement, and it refers to how evidence is secured from the time it was collected to the time it is used in decision making. A secure chain of custody means that access is controlled and that each individual with access is tracked. In research, chain of custody of data files is important for traceability. Recall that data from the analysis should be traceable back to the source. If there is a gap in the chain of custody, there is opportunity for data to become altered somewhere between the data capture and use, whereas if a secure chain of custody is in place, there is confidence that the data in the analyzed file reflect the data as they were collected.

Tasks Associated with Observation and Recording

Whether observation and recording are done manually or are automated, there are common steps to be taken into account in research planning. These include specifying observation procedures, making the observations, and recording them. Where there is any subjectivity, the observation procedures require testing and characterizing by measures such as inter- or intrarater reliability. Ongoing assessment is required to assure that the observation procedure performs as consistently as expected. Where instrumentation is used, calibration and ongoing assessment of calibration is required.

TABLE 6.5

Tasks Associated with Observation and Recording

Observation and Recording Tasks	All Cases	Special Cases
Specifying and documenting observation procedures	X	
Calibration of instrumentation and documentation thereof		X*
Testing and characterizing observation procedures including documentation thereof	X	
Making observations	X	
Recording observations	X	
Ongoing assessment and documentation thereof	X	

*When instrumentation are used.

The ongoing assessment activities should then be documented to demonstrate later that the observation and recording processes remained in control throughout the study. Table 6.5 enumerates the tasks associated with observation and recording. The table can be used as a checklist during study design, planning, and operationalization to assure that all aspects of observation and recording are considered and that an appropriate level of rigor is applied.

Summary

Data can be observed or measured manually by humans or in an automated manner by a device. These same options apply to data recording as well. Different sources of error apply to manual *versus* automated processes. Correspondingly, different methods for quality assurance and control are required. This chapter prepared the reader to recognize manual versus automated observation, measurement, and recording processes, and to select appropriate quality assurance and control measures for each.

Exercises

1. A civil engineer is testing three new varieties of concrete blocks to determine compressive strength in terms of maximum load under which the blocks will not crumble. He or she is conducting the experiment at three temperatures and four levels of humidity that are controlled in a laboratory testing chamber that can be set to different combinations of temperature and humidity. After increasing

the load in 100-pound increments, he or she visually inspects the surfaces of each blocks to detect and count cracks by visual inspection. The temperature and humidity values are provided by the testing chamber in the laboratory.

Which option below best describes the primary outcome measure?

a. Manual observation, manual recording
b. Manual observation, automated recording
c. Automated observation, manual recording
d. Automated observation, automated recording

Correct answer: (a) Manual observation, manual recording. Visual inspection is a manual process, and once the observation is made visually, the recording would be manual.

2. In the research scenario above, what actions could be taken to bolster confidence in the accuracy of the data?

a. Electron microscopy of the material surface to look for microcracks
b. Regular photographs of the material surface read by a second observer
c. Adding a second thermometer to the testing chamber
d. Having a second person review the crack count data

Correct answer: (b) Regular photographs of the material surface read by a second observer. The photographs preserve the event, and a second reader provides an opportunity to detect cracks missed by the initial observer.

3. In the research scenario above, which is the lowest cognitive load approach to data collection for the crack counts?

a. A table with the pounds in one column and the total number of cracks in the other column
b. Labels preprinted on the form for each poundage level with a blank for total crack count
c. A table for each block surface with the pounds in one column and the total number of cracks in the other column
d. A labeled photograph of each surface at each poundage level with new cracks circled on the photograph

Correct answer: (d) A photograph of each surface at each poundage level with new cracks circled on the photograph. This option is the lowest cognitive load because all the observer needs to do is to compare the photo to the before load photo to detect and circle the new cracks. Data recording tasks are reduced as much as possible.

4. If option d was chosen in the previous question, what would the researcher need to do during the tests?
   a. Count cracks
   b. Take photos and label by surface and poundage level
   c. Record crack counts in his or her laboratory notebook
   d. Take photos and label by surface, poundage level, temperature, and humidity

   Correct answer: (d) Take photos and label by surface, poundage level, temperature, and humidity. The parameters mentioned in option d need to be associated with the crack counts.

5. A cognitive psychologist is conducting usability tests on pharmacy labels affixed to prescription drugs prior to dispensing. He or she is interested in where people expect to see or find key pieces of information on a pill bottle label. He or she has created an 11 × 17 inch blank label and has cards for seven key pieces of information (drug name, dose, dosing instructions, prescription number, prescribing provider, precautions, and pharmacy phone number). He or she has three size cards of the same color for each piece of information. The 11 × 17 blank label is divided into six quadrants. To conduct the test, he or she sits down at a table with a research subject and asks him or her to place the information on the label as he or she would prefer to see it. He or she photographs the label created by the research subject after each session.

   Which option below best describes the observation and recording of the event of interest?
   a. Manual observation, manual recording
   b. Manual observation, automated recording
   c. Automated observation, manual recording
   d. Automated observation, automated recording

   Correct answer: (d) Automated observation, automated recording. The photograph is an automated observation and recording of the event (placement of the cards on the label).

6. In the research scenario above, which option best describes data collection about the event?
   a. Manual observation, manual recording
   b. Manual observation, automated recording
   c. Automated observation, manual recording
   d. Automated observation, automated recording

   Correct answer: (a) Manual observation, manual recording. To collect positional data about where on the 11 × 17 sheet, for example, where each card was placed and what size card was used, the researcher or designee has to manually review each photograph and record the location and size parameters.

7. In the research scenario above, what information must be associated with the location and card size data?
    a. The research subject
    b. The date, time, and research subject
    c. The observer, date, time, and research subject
    d. None of the above

    Correct answer: (a) The research subject. Association with the research subject should be maintained for traceability. If more than one observer were used, association with the observer would be needed to maintain attribution. Unless the data need to be associated with date or time, no additional associations are required.

8. In the question above why would the research subject need to be associated to the data?

    Correct answer: If the researcher wanted to see if demographic characteristics of the subject, for example, age, made a difference in placement or size, then the data (each photo) would need to be associated with the research subject and their characteristics. In addition, association of data with the research subject provides traceability.

9. A study is being conducted to test accuracy and patient acceptance of an implantable glucometer that measures blood glucose on the hour throughout the day. Each subject is given a device that connects via Bluetooth with the implanted device and sends a cellular signal to transmit the data to the electronic health record.

    List some potential sources of noise that might obscure the signal.

    Correct answer: Any of the following would be acceptable, the more complete the list the better, unless the student explains exclusions based on physics: cellular equipment such as phones, wireless Internet signals, electronics such as conventional and microwave ovens, and remote control toys. An understanding of signal transmission is not expected.

10. How might the study be improved?
    a. Each subject is given a standard glucometer that uses strips and capillary blood from a finger prick and asked to record blood glucose 1 hour before dinner.
    b. Each subject is asked to record daily symptoms associated with high and low blood sugar.
    c. Each subject is interviewed weekly about symptoms associated with high and low blood sugar.

   d. Each subject is given a standard glucometer that uses strips and capillary blood from a finger prick and asked to record blood glucose at the top of each hour during daytime hours.

   Correct answer: (d) Each subject is given a standard glucometer that uses strips and capillary blood from a finger prick and asked to record blood glucose at the top of each hour during daytime hours. A standardized independent instrument is the best assurance of accuracy. Recording blood glucose at the top of the hour while burdensome to participants, corresponds to the timing of the measurements taken by the implantable device.

11. In the blood glucose example above, which option best describes the observation and data recording from the implantable device for the study?
   a. Manual observation, manual recording
   b. Manual observation, automated recording
   c. Automated observation, manual recording
   d. Automated observation, automated recording

   Correct answer: (d) Automated observation, automated recording. The implantable device measures, records, and transmits blood glucose measurements.

12. In the blood glucose example above, which option best describes the observation and data recording from the handheld glucometer for the study?
   a. Manual observation, manual recording
   b. Manual observation, automated recording
   c. Automated observation, manual recording
   d. Automated observation, automated recording

   Correct answer: (c) Automated observation, manual recording. The handheld device measures blood glucose, but the research subject must record it.

References

Carmines EG and Zeller RA. *Reliability and Validity Assessment* (*Quantitative Applications in the Social Sciences*). Thousand Oaks, CA: Sage Publications, 1979.

Miller GA. The magical number seven, plus or minus two. *The Psychological Review* 1956; 63(2):81–97.

Shannon CE. Communication in the presence of noise. *Proceedings of the Institute of Radio Engineers* 1949; 37(1):10–21. Reprinted as a classic paper in: Proc. IEEE, vol. 86, no. 2 (Feb. 1998).

Wickens CD, Hollands JG, Banbury S, and Parasuraman R. *Engineering Psychology & Human Performance*. 4th ed. New York: Routledge, 2016.
Zhang J. A representational analysis of relational information displays. *International Journal of Human-Computer Studies* 1996; 45:59–74.
Zhang J and Norman DA. Representations in distributed cognitive tasks. *Cognitive Science* 1994; 81(1):87–122.
Zozus MN, Pieper C, Johnson CM, Johnson TR, Franklin A, Smith J, and Zhang J. Factors affecting accuracy of data abstracted from medical records. *PLoS One* 2015; 10(10):e0138649.

7

Good Data Recording Practices Applicable to Man and Machine

Introduction

Whether for a patent application, a research publication, or a regulatory decision making, the principles of good documentation practices are universal. This chapter covers good documentation practices with concentration on recording and preserving original data. Manual documentation strategies are discussed as are the use of instrumentation and computer systems, and situations that include both manual and computerized data recording.

Topics

- Necessity of good documentation
- Aspects of good documentation with respect to recording of data
  - Attributable
  - Legible
  - Contemporaneous
  - Original
  - Accurate

Necessity of Good Documentation

Recording data is the process of writing down or capturing data that have been observed or measured. Recording is the act of making a persistent record of the data, for example, copying down numbers from a visual readout or display, or using an instrument that creates an electronic record of the data. Recording applies to both the original capture of the data and the collection of existing data for a study. Arguably, every researcher has heard the three golden rules: (1) If you did not write it down, it did not happen; (2) if you did not write it down correctly, it did not happen; and (3) do not forget rules 1 and 2. This chapter is a brief description of what is required to appropriately record data for research.

As the chapter title indicates, this section concentrates on the aspects of research documentation relevant to recording of data. These are broadly referred to as *good documentation practices*. Students are often introduced to the principles of data recording in high-school biology or chemistry where they first use a laboratory notebook. The principles are taught again in various undergraduate laboratories. However, when students later work in a research facility, good documentation principles are not always emphasized. Good documentation practices are quite simple; they are the foundation for using research data to support decisions such as a patent application, a claim of first discovery, a regulatory decision making, a research publication, or a claim for or against scientific misconduct. Taking the lead of the United States Food and Drug Administration, the therapeutic product research and development industry has long used the terms attributable, legible, contemporaneous, original, and accurate expressed as the mnemonic *ALCOA* to represent the principles for data quality. ALCOA encompasses the main principles of good practices for data recording; thus, it is used as the organizing framework for the chapter. Woollen describes the regulatory basis for these principles with respect to data quality (Woollen 2010). These principles and related concepts are described below with examples of how they apply to the recording of research data. The European Medicines Agency adds Complete, Consistent, Enduring, and Available (CCEA) to the FDA's ALCOA principles (EMA 2010).

Attributable

Data values should be associated with the individual who or device which observed, recorded, and changed them as well as remain associated with the source of the data. In earlier examples, human observers have been described. Proper attribution in the case of a human observer is a formal association of the data value with the individual who observed and recorded it. A link in the data such as storing an observer identifier with the data is an example of a formal association. Where data are recorded in a laboratory notebook or field notes, this attribution is simple because researchers usually keep individual notebooks, and it is easy for an expert to assess the continuity of the writing. In cases such as large studies where there are multiple possible observers, observer initials or signatures are often captured with groups of data that are recorded together. A study signature log usually supports these identifying marks, so that each individual can be associated with his or her mark.

In electronic systems, this happens a bit differently. Some, unfortunately not all, systems require authentication of users, and the identity of the user is formally associated with any actions taken during the session. All changes to data are tracked in such systems. The tracking of data changes in this way is called an audit trail because it facilitates tracing back from a current value, through all data changes, to the original value. Further, some of the systems and organizations take additional measures such as system *validation*, *separation of duties*, and access restriction to assure that data cannot be entered or changed or updated without attribution. This concept is known as *nonrepudiation* and means that the system is designed and confirmed to operate in a way that users cannot deny accountability for actions taken under their user account. Specifics of nonrepudiation can be found in the three-page electronic record and electronic signature rule, Title 21 Code of Federal Regulations Part 11 (FDA 1997) (Table 7.1).

Attribution works similarly for changing recorded data. The accepted practice for changing data recorded on paper is to make one line through the value (s) to be changed, and to date and initial the change. The purpose of only one line through is so that the original value is not obscured. The date is for traceability so that the sequence of changes to data can be reconstructed and the most recent value is obvious. The purpose of the initial is so that the change is *attributable* to the person making the change. Some also document the reason for the change.

TABLE 7.1

Practices for Man and Machine

Man	Machine
Raw data must be recorded directly, promptly, legibly, and in permanent ink.	All values are recorded in an audit trail (preserves the initial entry).
Make original handwritten entries directly into a notebook, logbook, or appropriate study record.	The audit trail records the date and time of the change.
No pencil or erasable ink or water-soluble (usually felt tip) pens—they are not permanent.	The audit trail records the user identifier of the account making the change.
Ink should be indelible and dark enough to photocopy clearly.	The audit trail cannot be changed by users and cannot be bypassed.
Data should be signed and dated on the date of entry.	The system manual becomes part of the study documentation.
Data should be signed or initialed by the person making the entry. Sign and date records daily!	System testing and validation records become part of the study documentation.
Where there are loose pages, • Bind them or keep them in a file. • Label each page with sufficient information to uniquely identify the data and connect it to the study.	Data values for which the system is the original recording are documented.
Mark an X or line through the unused space.	Received data transfer files are preserved.
	Data transfer specifications become part of study documentation.

Legible

Legible simply means that the data are readable. Depending on the type of data entry or processing, some data, for example, handwritten recordings on labels or into a laboratory notebook, need to be read by humans. Other handwritten data, however, may be entered through a process called Intelligent Character Recognition (ICR); in such cases, legibility by the ICR software may require even neater writing as well as writing in the indicated areas on the form. Thus, legibility remains important. Computer recognition of marks and characters is covered in Chapter 8. Automated recording obviates concern about legibility. Recorded data that are not legible are essentially lost because they cannot be further processed.

Legibility is also a concern over time. For example, ink that fades would not be considered legible. Similarly, data that are archived are subject to loss if the media on which they are stored does not persist or the format in which the data are stored can only be read by computer systems that are no longer available. It is not unheard of for an author to receive a question about a paper, 10 years after the paper was published. Thus, legibility includes the notion that the data are preserved so that they remain legible over time. The European Medicines Agency separated out the latter concept and refer to it as enduring (EMA 2010).

Contemporaneous

It is expected that data are recorded at the time of observation or measurement. Recall from Chapter 5 that data recorded after the observation or measurement are subject to information loss due to the passage of time. *Contemporaneous* data recording also corroborates the sequence of observations or measurements.

The concept of contemporaneity can be extended to the identification and resolution of data discrepancies. In general, it is accepted best practice in time to screen data to identify and resolve discrepant values as close as possible to the original recording or collection of data for a research study. The practice has been historically justified by the cost of rework. For example, the *one ten–one hundred rule* states that if it costs $1 to identify and resolve a discrepant data value at its origin, it costs $10 to identify and correct the same data value during data processing because one has to go back and consult the source or original recording; it costs $100 to identify and resolve discrepant data during later stages of a project. The recommendation to identify and correct data discrepancies as soon as possible is also justified from a data quality standpoint. As described in

Chapter 5, the passing of time invites information loss and degradation, and in some cases renders data uncorrectable. Thus, data should not only be recorded contemporaneous with the observation or measurement but also be cleaned at that time if at all possible.

Original

The intent is that the data for a study are traceable back to the *original recording*. With electronically observed and recorded data, the data are the original recording. With data recorded on paper, as we have seen in Chapter 5, this may not be the case. There are plenty of research scenarios where the collected data are not the original recording, for example, data abstracted from veterinary records and entered onto a data collection form. Here, the veterinary record is the original recording of the data; *abstracting* and *transcribing* the values onto the data collection form is the data collection (recording) of the data for the research study. In these cases, it is important for the researcher to identify which data are the original recording and which are not, and for the data that are not the original recording, the researcher should document the source of the data. The latter provides traceability, in this case a path from the data back to the source. If necessary, the data can be compared back to the source to verify that the data are free from adulteration (Table 7.2).

TABLE 7.2

Dos and Don'ts of Data Recording

Dos	Don'ts (Preventing Common Data Recording Mistakes)
Do record complete data	Do not obliterate or obscure an original value
Do correct mistakes	Do not write the corrected entry over the original value
Do crossout incorrect values with a single line through	Do not use light-colored ink
Do date and initial every correction	Do not use water-soluble pens (many felt tip pens)
Do include reason for change where not documented elsewhere	Do not leave blank space
Do date and sign each page of handwritten unbound information	Do not pre- or postdate entries
Do use a nonambiguous date format	Do not discard original records
Do use military time, or designate a.m. or p.m.	
Do line through blank space or indicate intentionally left blank	

Accurate

Accurate data reflect the truth at some stated or implied point in time. Data accuracy is the property exhibited by a data value when it reflects the true state of the world at the stated or implied time point. It follows then that an inaccurate data value does not reflect the true state of the world at the stated or implied time point. *Data errors* are instances of inaccuracy, and data *discrepancies* are suspected or possible data errors.

Data accuracy is easy to define but difficult to directly measure unless one has a source of truth against which to compare data. For this reason, surrogate measures are often used for data accuracy. Surrogate measures do not measure accuracy directly, but measure other things that may indicate accuracy. Examples of surrogates include the percentage of data values that contain logical inconsistencies (discrepancies) compared with other data values or the percentage of data values that are discrepant compared with some independent source of the same information such as a split sample sent to two different laboratories. These examples are measures of consistency used as surrogates for measuring accuracy. They all involve making comparisons in hopes of identifying discrepancies. The closer to truth the comparator is, the closer to accuracy the surrogate measure is (Zozus et al. 2014). Thus, accuracy can be conceptualized as a type of and the ultimate consistency; consistency with the truth.

Data accuracy has been described in terms of two basic concepts: (1) representational adequacy or inadequacy, defined as the extent to which an operationalization is consistent with or differs from the desired concept (validity), including but not limited to imprecision or semantic variability, hampering interpretation of data, and (2) information loss and degradation, including but not limited to reliability, change over time, and error (Tcheng et al. 2010).

Representational inadequacy is a property of the data element. It is the degree to which a data element differs from the desired concept. For example, a researcher seeking obese patients for a study uses Body Mass Index (BMI) as the operational definition of obesity, knowing that a small percentage of bulky but lean bodybuilders may be included. Representational inadequacy is best addressed at the point in research design when data elements and sources are selected. (Operationalization and concept validity is fully described in Chapter 3.)

Information loss or degradation is the reduction in information content over time and can arise from errors or decisions in data collection and processing (e.g., data reduction such as interval data collected as ordinal data, separation of data values from contextual data elements, or data values that lose accuracy or relevance over time). Information loss and degradation may be prevented or mitigated by decisions made during research design. Such errors and omissions are sensitive to factors such as local or study documentation practices and *mapping* decisions made for data in the design of data collection forms. Information reduction is addressed in Chapter 3 and information loss is addressed in Chapter 5.

Accuracy is somewhat different from the other ALCOA components because although, like the other components, a researcher designs for accuracy to be present, however, the actual degree of accuracy achieved should be measured. Measurement is necessary for early detection and correction (control and management activities) and also at the end of a study to demonstrate that the data are capable of supporting research conclusions.

Completeness of research data is often taken to mean that all of the planned data were collected. Conducting a completeness review too late in the research process may mean that the opportunity to obtain the data is lost (Chapter 5). Thus, in research contexts the general expectation is that data are reviewed as early as possible in the data collection process (1) to assure that the needed data were provided, and (2) to take corrective action if data are missing to obtain the data where possible or to improve the data collection process or both. Such a review provides feedback to data collectors and enforces the need for all of the data. Such a review also will identify data collection procedures that may not be feasible or may fall victim to unanticipated operational challenges so that they can be corrected or mitigated as early as possible.

Availability of data is a concern during all phases of a research project. The planning stage should identify data availability needs during the project, for example, if early stopping rules are used, proper execution of the project relies on clean data for one or more interim analysis. Further, large or complex projects often rely on access to data to track the progress, for example randomization balance, or to report project status. These during project data uses must be planned for such that the data collection and processing supports timely availability of data. Data availability is also necessary after publication of the results and subsequent archival, for example for data sharing or to answer questions about the research.

An Urban Legend about Data Recording

Where does the *black ink* requirement come from? Older copy machines needed the high contrast of black ink on white paper to make high-quality copies. This is not a factor with today's copy machines, but black ink is often still specified.

Summary

This chapter covered good documentation practices for both humans and machines. The necessity of good documentation was described, and nine aspects of good documentation were presented. The focus was on good documentation in the recording and updating of data. Documentation of data management-planned activities was covered in Chapter 4, and research documentation will be covered holistically in Chapter 13.

Exercises

1. A manufacturing facility is running some experiments in support of quality improvement. The line workers measure one randomly selected widget per hour to see if the piece conforms to the specified tolerances for length and height dimensions. At the end of the shift, the line workers record the pass/fail information for the eight widgets measured on their shift. Indicate which of the following ALCOA principles are violated:
   a. Attributability
   b. Legibility
   c. Contemporaneity
   d. Originality
   e. Accuracy
   f. None of the above

   Correct answer: (c) Contemporaneity. By recording data at the end of the shift, the line workers are not recording it contemporaneously with the observation.
2. An internal auditor for a research study finds several instances where the data do not match the original records in the researcher's field notes. Indicate which of the following ALCOA principles are violated:
   a. Attributability
   b. Legibility
   c. Contemporaneity
   d. Originality
   e. Accuracy
   f. None of the above

   Correct answer: (e) Accuracy. Although we do not know that the data fail to reflect the true state of the observed phenomena, the auditor detected discrepancies and discrepancies are often used as a surrogate for accuracy, so accuracy is the most appropriate choice.

3. A researcher receives a question about a paper for which a post-doc in his or her laboratory was the first author (the researcher was the corresponding author). The question is received 12 years after the paper was published. The researcher gets the study documents including the laboratory notebook back from the archival facility. Unfortunately, the values recorded in the laboratory notebook have faded such that several of them are not readable. Indicate which of the following ALCOA principles are violated:
   a. Attributability
   b. Legibility
   c. Contemporaneity
   d. Originality
   e. Accuracy
   f. None of the above

   Correct answer: (b) Legibility. The faded ink is an example of illegibility.
4. In the question above, the researcher uses the SAS data sets from the study analysis to answer the question because they were easily readable in the current version of SAS. Indicate which of the following ALCOA principles are violated:
   a. Attributability
   b. Legibility
   c. Contemporaneity
   d. Originality
   e. Accuracy
   f. None of the above

   Correct answer: (d) Originality. The original recording of the data could not be used.
5. A member of the research team is checking the analysis by verifying counts in cells on one of the tables using the data from the data collection forms. For this study, the data collection forms happen to be the original recording of the data; from the data collection forms, the data were entered into a database and then analyzed to create the tables. He or she is unable to replicate the counts. Indicate which of the following is the best label for the problem:
   a. Lack of attribution
   b. Data entry or programming errors

c. Illegible data
d. Lack of traceability

Correct answer: (d) Lack of traceability. We cannot tell from the description if the inability to replicate the counts was caused by errors. All we know is that the numbers in the cells on the table could not be traced back to the raw data.

6. In the question above, which of the following could be questioned as a result of the data discrepancies identified?
   a. Attributability
   b. Legibility
   c. Contemporaneity
   d. Originality
   e. Accuracy
   f. None of the above

   Correct answer: (e) Accuracy. Although data discrepancies are not a direct measure of accuracy, their presence often indicates accuracy problems.
7. In a small research study, a researcher had tree girth data entered directly into a spreadsheet. There was a column for each of the following data elements: lot number, tree identifier, tree girth at four and a half feet, and date of measurement. The tree girth measurements were taken in inches, and a calculated column was added for the girth in meters for a total of five columns. Which of the following have been violated?
   a. Attributability
   b. Legibility
   c. Contemporaneity
   d. Originality
   e. Accuracy
   f. None of the above

   Correct answer: (a) Attributability. There was no column to indicate the observer. In addition, there were no columns to indicate who added the converted girth, or when it was done.
8. In the question above, what might also be said about the data?

   Correct answer: Without a mechanism to track changes to the data, the data lack traceability. Also, there is no way to assess the accuracy of the data, and as the scenario was described, no steps taken to assure data accuracy.
9. In question 7, which of the following would be required for the researcher to be able to make a claim of nonrepudiation?

a. Using a system that tracked reason for change and person making the change
b. Using a system that required users to authenticate themselves
c. Associating user ID with entries made under the account
d. Validating that the system works correctly
e. All of the above
f. All of the above and more
g. None of the above

Correct answer: (f) All of the above and more. The additional things include preserving the data values prior to the change, tracking the date of the change, and controls, so that the audit train cannot be bypassed.

10. An auditor reviews the data collection forms for a study and notes a finding of undocumented changes to data. Which of the following would have prevented this finding?
a. Reviewing the data to identify and correct undocumented changes
b. Using a system that tracks all changes to data
c. Training all data collectors in good documentation practices
d. All of the above
e. (a) and (b) only
f. (a) and (c) only
g. None of the above

Correct answer: (e) (a) and (b) only. Training all data collectors can help but will not prevent human error. (a) is included in the correct options because with paper processes, (a) is the only way to reduce the likelihood of—but not prevent entirely—such findings.

References

DEPARTMENT OF HEALTH AND HUMAN SERVICES, FOOD AND DRUG ADMINISTRATION, TITLE 21 Code of Federal Regulations Part 11, Electronic records; electronic signatures. available from http://www.accessdata.fda.gov/scripts/cdrh/cfdocs/cfcfr/cfrsearch.cfm?cfrpart=11. Accessed March 12, 2017.

European Medicines Agency, Reflection paper on expectations for electronic source data and data transcribed to electronic data collection tools in clinical trials. EMA/INS/GCP 454280/2010, June 9, 2010. Accessed on March 12, 2017, available from http://www.ema.europa.eu/docs/en_GB/document_library/Regulatory_and_procedural_guideline/2010/08/WC500095754.pdf.

Tcheng J, Nahm M, and Fendt K. Data quality issues and the electronic health record. *Drug Information Association Global Forum* 2010; 2:36–40.

Woollen SW. Data Quality and the Origin of ALCOA. The Compass—Summer 2010. Newsletter of the Southern Regional Chapter Society or Quality Assurance, 2010. Accessed November 17, 2015. Available from http://www.southernsqa.org/newsletters/Summer10.DataQuality.pdf.

Zozus MN, Hammond WE, Green BG, Kahn MG, Richesson RL, Rusincovitch SA, Simon GE, and Smerek MM. Assessing Data Quality for Healthcare Systems Data Used in Clinical Research (Version 1.0): An NIH Health Care Systems Research Collaboratory Phenotypes, Data Standards, and Data Quality Core White Paper. 2014. Accessed November 26, 2016. Available from www.nihcollaboratory.org.

8

Getting Data into Electronic Format

Introduction

For all but the smallest of data sets, a computer is needed for data processing, storage, and analysis. As the first data processing step, the data usually must make their way into electronic format where they can be read and operated on by computers. There are multiple ways of getting data into electronic format. These methods span the continuum from completely manual to fully automated, with the majority of methods and associated tools belonging to the semiautomated variety. Although all methods are important, as different situations may favor one over others, researchers should pay special attention to and seek direct electronic data acquisition from devices. Everyday more data become available from electronic measuring devices, and the cost and barriers to use of such devices continue to decrease.

This chapter presents manual, semiautomated, and fully automated methods for getting data into electronic format. The methods covered include key entry, computer recognition, and direct electronic capture. Common applications, advantages, and disadvantages of each are discussed, as are special considerations for structured, semistructured, and unstructured data. This chapter prepares the reader to choose and apply data entry methods and to prospectively identify potential sources of error and appropriate prevention or mitigation strategies.

Topics

- Preprocessing
- Getting data into electronic (computable) format
  - Mark and character recognition methods
  - Voice recognition

    - Keyboard entry
    - Direct capture from devices
- Standardizing manual processes

Getting Data into Electronic Format

For data to be easily used, analyzed, or shared, they need to be electronically available. For example, values written on a piece of paper usually need to be typed into a calculator or other computer before mathematical operations can be performed on them. Electronic format in this context means that data are machine readable or usable by computers.

Consider a benchtop experiment to assess the effectiveness of two different cell growth media. Information about the care of cells, the status of medium changes (refreshes), cell counts, and other observations may be recorded in a laboratory notebook. If the duration of the research, the number of data points collected, and the number of cultures are small, the data could reside in a properly maintained laboratory notebook for the duration of the study. At the end of the experiment, data may be entered into a calculator, a spreadsheet, or statistical software for analysis.

In this scenario, the data processing tasks are straightforward. If the analysis is done on a calculator, the analysis should be repeated to guard against typographical errors. Other than assuring the initial measurement or observation and recording, good documentation practices, data entry quality (in this case by repeating the analysis calculation), and the researcher's ability to detect data errors in a small data set, nothing else is needed. Alternatively, consider the same experiment but with hundreds of samples and additional data to be collected. At a minimum, a spreadsheet, and potentially a database would be a better choice for processing, management, and storage of data. In addition, if many values are hand entered, a quality control on the entry is necessary to assure errors were not generated during the data entry process and any subsequent processing of the data. The same researcher may opt for a system where data are recorded on paper data collection forms and scanned into the computer where the form images are processed by optical mark recognition (OMR) or intelligent character recognition (ICR) software. Some of the data may be available electronically from other information systems, for example, those associated with laboratory instrumentation, and captured electronically initially; in this case, there is no data entry step. Additional examples of direct capture from devices were covered in Chapter 6.

Recall from Chapter 5 that delays in data collection and cleaning decrease the achievable data quality. Thus, data should be converted into electronic form as soon as possible. Data on sheets of paper cannot be easily searched for discrepancies. In general, data on paper are more difficult to process due to the manual operations usually required. Without data in electronic form, a manual process must be set up to identify, track, find missing pages, and mark them off of the list as they are received. A similar process must be set up to manually review each page of data and identify discrepancies. With a small study, this is possible to do by hand, but such manual processes are not feasible on a large study. Manual processes are also error prone. The published human inspection rate is about 15% (Anderson and Wortman 1998). This means that a human looking at data to identify discrepancies will miss about 15% of them. The most efficient and accurate way to manage large amounts of data is to get the data into electronic form as soon as possible. The total number of data values for a study can be easily calculated as the number of experimental or observational units multiplied by the number of data values collected per experimental or observational unit. For example, a study with 50 observational units and 25 data values collected per unit would have 1250 total data values expected. The lager this number, the less feasible it is to manage data manually.

Methods to get data into electronic formats where they can be operated on by computers include key entry, optical scanning methods, voice recognition, and instrumentation-based direct electronic capture. This is not an exhaustive list and new technology is being developed every day (Figure 8.1).

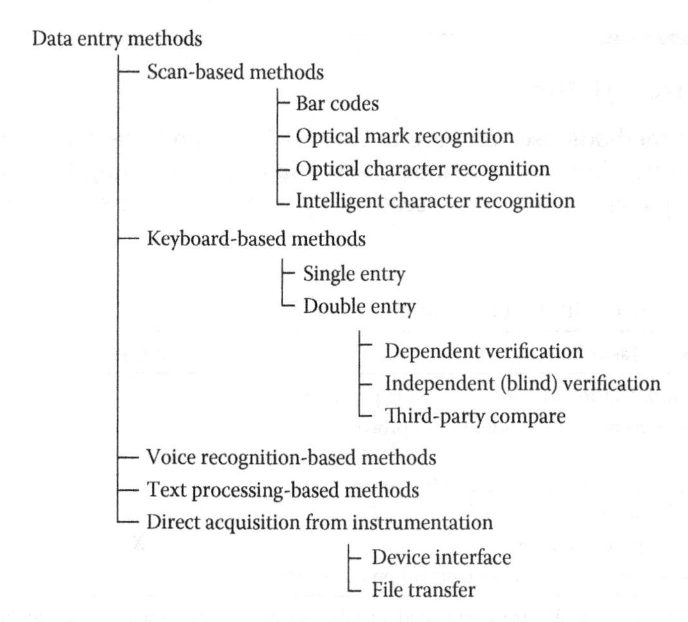

FIGURE 8.1
Data entry methods.

Preprocessing

There are often things that must be done to get data ready for entry. Examples include checking that all expected pages are present for multipage forms and checking to see that all pages (or items otherwise received separately) are appropriately labeled with the identifier for the experimental or observational unit and time point. These activities are called *preprocessing* activities (Table 8.1) are often but not always performed manually before data are in electronic format. When preprocessing steps are manually performed, they should only include things that would prevent further processing of data, for example, checking incoming data files to assure that the experimental unit identifier is included in each file. As many of the data checking tasks as possible should be done as soon as data are in electronic format. There is a trade-off between checking for discrepancies as early as possible in the process (some preprocessing tasks) and checking for them at the point in the processing where such checking is most efficient.

Tasks associated with preprocessing are enumerated in Table 8.1. Preprocessing that includes manual steps requires initial assessment with some measure of reliability and ongoing assessment of consistency to demonstrate that the procedures perform as expected. Preprocessing via automated means requires some programming or other method of setting up the algorithm as well as initial testing, validation and ongoing monitoring to demonstrate that the algorithms and their implementation perform as expected.

Scan-Based Methods

Scanning methods use software to read and interpret the marks on paper, and write the data into a database essentially transforming the data from marks on paper to a format processable by computers. The scanning process

TABLE 8.1

Tasks Associated with Preprocessing*

Preprocessing Tasks	All Cases	Special Cases
Specifying and documenting preprocessing procedures	X	
Programming, testing and validating preprocessing algorithms		X
Testing and characterizing preprocessing procedures including documentation thereof	X	
Performing preprocessing tasks	X	
Ongoing assessment and documentation of preprocessing tasks	X	

*Most preprocessing tasks are performed in all cases where preprocessing is needed—those marked "All Cases". In special cases, data are available and received electronically, in these cases, preprocesing tasks can and should be programmed.

is conceptually similar for the different types of scanning technologies. The forms are scanned (a fax machine can also be used as a scanner). Software called a *recognition engine* reads the information from the scanned image and converts it to machine-readable data. Many recognition engines usually provide a confidence or likelihood of correctness for each value. In this case, there is usually a postscan verification step where the scanned data below a configurable confidence threshold is displayed on a computer screen for a human operator to confirm and *verify* the information.

There are several different types of scanning technology used today, including one-dimensional (lines) and two-dimensional (boxes) bar codes, *magnetic ink character recognition* (MICR), *optical character recognition* (OCR), OMR, and ICR. There are some specialized applications where scanning technology has become common. Survey research where individuals are mailed or given a questionnaire to complete has historically been one such application. A paper form is completed and submitted to the researcher or a data center for processing. The key factor making survey-based research a good candidate for scanning technology is that discrepancies are not resolved after the fact by contacting the individuals completing the questionnaire. In older paper-based survey models, discrepancy identification and resolution processes other than that by the recognition engine to increase recognition fidelity was not necessary. Although many surveys today are implemented electronically where discrepancy checking can occur real time during form completion, there are still applications where paper is used, usually cases where access to potential respondents is limited, or in settings where completion of a paper-based form is easiest. A similar situation occurs in behavioral research and quality-of-life assessment. Where the data come from validated questionnaires that are completed directly by the research subject. Like survey-based research, the respondents are not contacted after the fact to resolve data discrepancies. "How did you feel 2 weeks ago?" is not a valid question when the questionnaire was designed to collect how a respondent feels at the time they complete the form. Workflow is to hand respondents a questionnaire, have them complete it, and send it to the data center for processing. A final example occurs in remote parts of the world where cellular and internet connectivity is sparse or expensive. Where large volumes of data are collected in remote regions, a scan-based process have historically often been the most feasible.

Bar Codes

Bar codes are used in three common ways: (1) preprinted on a document as a page or document identifier, (2) preprinted on separator pages indicating the beginning or end of a document, and (3) preprinted or printed on demand labels to be affixed to documents or objects to track them. They are often used to automate workflow, especially tracking items through a workflow.

For example, when a central laboratory is used for a study, barcoded labels are often affixed to sample containers. The bar-coded samples are sent to the laboratory with a laboratory sheet on which some additional data are provided. With the investment of a $10 bar code reader, the average personal computer can read bar code data in a fraction of a second with almost no errors. Many programs can generate bar codes or bar coded labels.

There are a number of different types of bar codes called *symbologies*. Examples of popular one-dimensional symbologies include Code 39, Interleaved 2 of 5, and Postnet. One-dimensional symbologies consist just of lines and can hold only a very limited amount of information. Depending on the size of the code and the symbology, there may only be room for 12–24 characters. There are also two-dimensional symbologies that encode hundreds of characters. PDF 417 is the most popular two-dimensional symbology. The interested reader is referred to *The Barcode Book* by Roger C. Palmer (2007).

Barcodes have disadvantages. For example, if a bar code becomes damaged and the line pattern is obscured, it may not be readable by the recognition engine. It is important to always have a human-readable version of the value on the same page as the bar code itself. Many bar code packages include this option. The disadvantage of using two-dimensional bar codes is that normally there is also no room to print the human-readable version of the encoded data. That means that a failure to read a PDF 417 code will result in the inability to lift data from the bar code. PDF 417 and other two-dimensional barcodes include redundancy in their values to help prevent damage to the barcode from obscuring the encoded information. Table 8.2 enumerates the tasks associated with bar code use.

Use of bar codes requires some tasks specific to bar coding. First, the data elements to be bar coded need to be selected. In research contexts, these are usually unique identifiers for experimental or observational units, time point identifiers, or sample identifiers. Thus, the association of bar coded fields with other data is specified early. From a workflow perspective, the method by

TABLE 8.2

Tasks Associated with Bar Code Use

Barcoding Tasks	All Cases	Special Cases
Selecting data elements for bar coding	X	
Specifying association of bar coded data with other data	X	
Selecting a bar code standard, hardware, and software	X	
Selecting and specifying method for bar code printing		
Specifying and documenting bar coding procedure	X	
Installation of bar coding hardware and software	X	
Testing and characterizing bar code reading including documentation thereof	X	
Performing bar code reading tasks	X	
Ongoing assessment of accuracy and documentation thereof	X	

which bar codes are printed and affixed to forms, samples, or other artifacts is critical. For example, where bar codes are used as time point or assessment identifiers, they have often been printed directly on data collection forms, or provided as sticker labels accompanied by sample containers to which they are affixed before submission. The bar coding procedure and workflow must be specified and documented. Given the fields to be bar coded, the bar code standard, hardware, and software are selected. Once all of this is completed and bar coding software and hardware is installed, the bar coding process is tested and characterized, and this validation is documented. Once the study starts, bar code use and reading occurs throughout the study. Periodically, the process is assessed to confirm that it is performing at the expected level and with the expected consistency. There is little variability in this high-level set of tasks associated with barcode use; all tasks are usually performed in all cases.

OMR and OCR

Recognition engines are specialized for marks, machine-printed characters, or handwritten characters, and some software packages include multiple types of engines. Data collection forms with fill-in bubbles or check boxes can be read with, and by, an *OMR* engine. OMR is used in several research applications.

During development, OMR software is provided thousands of samples of marked boxes, circles, and filled bubbles to train the software. Many samples are required because human marking of bubbles or boxes can be quite variable. Check boxes can be marked with an x, a check, or a line through in various directions. When the software scans, it uses the training information to calculate the confidence in the accuracy of the scanned value for each bubble or box. The software is usually configured by setting a confidence threshold such that boxes or bubbles below the threshold will be flagged for verification by a human. The values marked for verification usually appear in a different color or with some other visual distinction on the screen to alert the operator that verification is required. OMR software often has trouble with common form completion problems such as forms completed by drawing a line across a series of boxes to indicate that all are selected. The line may wander in and out of individual boxes, and check-mark tails may stray over part of another box.

Optical Character Recognition, or *OCR* reads machine print, or typed information, not handwritten characters. It can be utilized in research where printed documents are available but not in electronic form. For example, a researcher is conducting a systematic review and some of the articles are older and available only as images and not text. He or she is using a qualitative analysis software package to code the study designs in the manuscripts included in the review. The qualitative analysis software allows him or her to import documents, but will only allow him or her to mark passages and code them for the newer manuscripts. Passages from the older manuscripts available only as image files cannot be coded in the qualitative analysis software. The researcher uses an OCR engine to convert the older manuscripts to text files that can be read and coded in the qualitative analysis software. OCR can be used whenever machine-printed material is available to render the text processable.

Intelligent Character Recognition, or *ICR*, is used to recognize hand-printed characters. Some ICR engines will read machine print, but are usually not as accurate as a specialized OCR engine for machine print. Form design and layout can dramatically impact the accuracy of ICR data extraction. For example, if data fields are too close to the edge too densely packed together, or not well formed, characters will be mistaken (Figure 8.2). Recognition is best when the characters do not touch each other. For this reason, boxes are printed on the forms in a *drop out* or *sterile* color that is not seen by the scanner. The boxes help the individual completing the form to neatly place the characters in the boxes and provide a measure of character separation needed by the engine. ICR works best on neat handwriting. Like OMR, ICR software can be configured with a threshold confidence so that items of questionable accuracy can be verified by a human.

Though OMR and ICR methodology automate the entry of data, forms with bubbles and boxes take longer for the respondent to complete. Another disadvantage is that manual work is not eliminated; the verification step to confirm (or not) values flagged for low confidence requires a human to confirm or correct the value. A human may also be needed to screen forms before scanning for proper orientation and to associate the form with the experimental or observational unit, the time point for the measurement, or other relevant context. In addition, pages must be handled carefully because damaged or misaligned pages get stuck or skewed in scanners. Where forms are scanned by facsimile, the sender has to be contacted and has to resend pages that are skewed or not received. When a response is marked on an OMR form, the value associated with the box must be mapped to a specific location in the database. This mapping is done after the form is designed and before the system can be used to scan data, and requires time as does testing and configuration of the confidence threshold level(s). Owing to the fact that form completion and mark

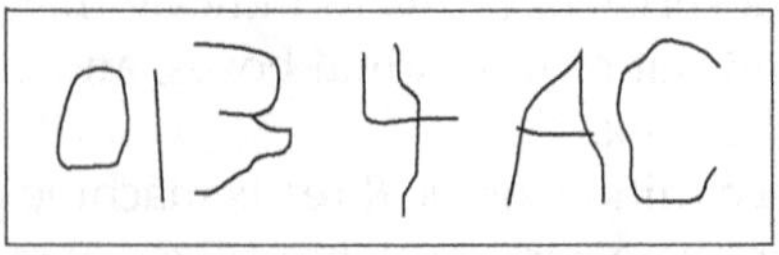

Is the correct value 934AC or one of the following?
0134AC
OB4AC
9B4AC
OB4AC

FIGURE 8.2
Challenges faced by an ICR engine.

or character recognition is sensitive to the design and layout of the form, the software must be tested on forms for the research study, and the confidence threshold set for each study. Such testing should be confirmed by scanning error rates measured during the study and reported with the study results.

From a workflow perspective, OCR, OMR, and ICR are similar. The scanning procedure must be specified and documented. Scanning software and hardware must be selected and installed. An initial system validation should take place. Often scanning systems come with an example form and test data already set up. Initial testing should confirm that the system operates as expected with the test form. This is sometimes called out-of-the-box testing. For a given study, the data collection form needs to be designed. Form design is often facilitated by the software, as is the mapping of the fields on the form to data storage structures. Scanning confidence thresholds need to be set for the study forms, and data discrepancy checks for missing, out-of-range, and logically inconsistent data are programmed or otherwise configured. After all of this is completed, the results are used to set the scanning thresholds. During the conduct of the study, scanning, verification, and discrepancy resolution tasks are performed. The time for each can be estimated during the study-specific testing. Ongoing assessment and documentation thereof occurs throughout the study (Table 8.3). Two of the tasks are indicated as performed in special cases. Mapping of forms to data storage structures is often done automatically by the software as part of form design; where this is not the case, the mapping step must be added. Programming or other set-up of computerized checks to identify data discrepancies is recommended and usually needed; it is marked as a special case because many OMR, OCR, and ICR software packages do not include this functionality and setting up discrepancy checks becomes an additional task or must be performed external to the software.

TABLE 8.3

Tasks Associated with Optical Mark and Character Recognition Based Entry Methods

Scanning Tasks	All Cases	Special Cases
Specifying and documenting the scanning procedure	X	
Selecting scanning software and hardware	X	
Installation of scanning hardware and software	X	
Initial system validation (out of the box testing)	X	
Designing the data collection form	X	
Mapping of the form to data storage structures		X
Programming or otherwise setting up data discrepancy checks		X
Setting scanning confidence thresholds for the study	X	
Testing and characterizing study-specific scanning procedures including documentation thereof	X	
Preprocessing of forms for manual scanning	X	
Performing scanning and verification tasks	X	
Ongoing assessment of accuracy and documentation thereof	X	

Speech Recognition

Speech recognition is the parsing and translation of spoken language into text by computers. It is also called automatic speech recognition, computer speech recognition, or *speech-to-text* processing when done by computers. Transcription is a speech-to-text processing task performed by humans. Speech recognition services, for example, transcription services and software, are readily available today, but there are few designed to support researchers. Such systems have been used widely in clinical trials to support randomization of research subjects and to collect patient-reported data via the telephone. In both of these applications, the vocabulary to be processed is relatively small, usually yes or no responses or numbers. A combination of key entry and voice recognition via the telephone is often supported. These systems have in common that the number of data values collected is low, such that data are entered and collected in a quick call. Using telephone-based speech recognition for a large number of questions grows tiresome for the respondent because such systems are slow compared to other ways humans transmit information. Speech recognition systems basically convert the spoken work to machine-readable character and numeric data.

Different types of broader voice recognition systems have been developed to support professionals in varying disciplines, but not yet a general research application. Where professional systems or general speech recognition applications are used, the researcher bears the burden of associating important context to the data files, for example, appropriately associating the data files with the appropriate experimental or observational unit and time point. For example, in a community research study, telephone interviews were recorded by a commercial call recording vendor. The vendor was able to make files available via a secure internet site, and the files were labeled with the date and time of the call. Associating the files with the correct research participant had to be done by stating the participant number at the start of the call and editing the filename to include the participant number. Note in the example above electronic files (readable by a word processor) were provided containing the text. Additional functionality would have been needed to associate recognized words or phrases with meaning or data elements, that is, to further convert the text files to structured data. For a detailed description of the use of voice recognition in animal behavioral research, see White et al. (2002).

Tasks involved in use of speech recognition systems (Table 8.4) include those for selecting, installing, and testing software. In addition, some software may include the ability to train the recognition engine or to set confidence levels. First, the researcher must specify the data to be collected via vocalized responses from observers or research participants as well as the workflow. In addition, the researcher must select transcription resources or voice recognition software. Recognition hardware and software must then be acquired, installed, and subjected to initial system validation. If the

TABLE 8.4

Tasks Associated with Transcription or Automated Speech Recognition

Recognition-Related Tasks	All Cases	Special Cases
Specifying data to be collected via vocalized responses from research team members or research participants	X	
Selecting transcription resources or voice recognition software and hardware	X	
Specifying and documenting the recognition procedure	X	
Installing recognition software and hardware		X
Initial system validation (out-of-the-box testing)		X
Mapping data to data storage structures		X
Setting recognition confidence thresholds	X	
Setting up manual review process for items falling below the low confidence threshold	X	
Testing and characterizing study-specific recognition procedures including documentation thereof	X	
Performing recognition-related tasks	X	
Ongoing assessment of accuracy and documentation thereof	X	

functionality is available, the recognition engine may then be trained on study-specific vocabulary. A specific study is set up in the system, and collected data elements are mapped data to data storage structures or output files, and recognition confidence thresholds are set. Logic such as changes in question order based on responses to prior questions are added during study configuration. After the study is configured or otherwise set up in the system, study-specific recognition procedures are tested, characterized, and documented. Recognition-related tasks are performed throughout the study conduct, and the reliability is assessed and documented periodically throughout the study. This sequence of steps is deceptively simple given the variety of ways in which speech recognition methods are used in research. These range from use of humans to transcribe or code selected items from recorded voice to computerized transcription, to computerized systems that ask questions, and employ speech-to-text processing to extract structured data from the responses. Often, text processing algorithms are specialized for a particular application such as extracting structured data from radiology reports, or as described in White et al. (2002) from a monolog of experimental observations.

Key Entry

Keyboard-based data entry is a process in which a human called a data entry operator reads data from paper or other image and types the data into a computer. There are four main types of key entry: (1) *single entry,* (2) *double entry with dependent verification,* (3) *double entry with independent verification,* and (4) *double data entry with third-party compare.*

Single entry is a simple process. Data are recorded on a paper form. That form is then later entered. A data entry operator keys data on the form into a database system, a statistical analysis package, or a spreadsheet. Data quality from single entry depends on the process used. Single entry with no on-screen checks has been associated with error rates close to 100 errors per 10,000 fields. Adding on-screen checks to single entry has been associated with lower error rates in the range of 20 errors per 10,000 fields (Nahm 2012). The largest reduction will likely be seen with comprehensive on-screen checks for data entered in the field—recall the timing of key milestones in Chapter 5. The amount of error reduction obtainable from on-screen checks is completely dependent on the type of checks that are written and how well they cover the error sources.

Single entry requires more attention than double entry. If double entry operators are used for single entry without training and acclimation time, the error rate will be higher. This is because speed is emphasized in double data entry, and double data entry operators are used to having a safety net of a second person entering the same data. There are of course plenty of cases in the published literature of good quality from single entry where data were entered with assurance and control mechanisms in place such as entry operators self-checking their own or each others work or some other type of regular review and feedback.

Double data entry is a process in which data are entered twice, usually by two different individuals. The purpose for double data entry is to catch and correct keystroke errors during the data entry process and as early in the data processing stream as possible. In double entry, the data are collected on a paper form or transcribed from some prior recording onto paper forms. The forms are often entered into a computer system in a central location such as the researcher's office or a data center. At the data center, a data entry operator keys the data into a computer system. Then the same data are entered the second time. The first and second entry values are compared to identify discrepancies. The discrepancies are corrected before the data move to the next stage of data processing.

The error rate from double data entry ranges from a few to 20 errors per 10,000 fields (Nahm 2012). Double entry is the most efficient way to eliminate keying errors. Double entry is most effective when the first entry operator is a different person than the second entry operator. There are three basic variations on the double data entry process:

Third-person compare: The first and second entry data are compared, usually electronically, and a third person, different from the first and second entry operators, reviews the list of differences, decides the correct value, and makes corrections.

Interactive verification: Here, the first and second entry values are compared by the computer system as data are entered, with the second entry operator addressing differences, deciding the correct value, and making corrections during the second entry. There are two types of interactive verification: (1) *blind verification,* in which the second entry operator is not aware of the value entered by the first entry operator, and (2) unblind or *dependent verification*, in which the second entry operator sees the actual value entered by the first entry operator. Blind interactive verification seems to be the process used most often. In interactive verification, the more senior operator performs second entry.

Use of key entry in a research study starts with specifying and documenting the data entry procedure in data entry guidelines. Data entry guidelines state instructions for how to enter each field and pay special attention to the potential variation in the incoming data. For example, field-specific guidelines for numerical data where form fillers have entered a range instead of a single number will give clear instruction on whether the value should be left blank, should be resolved with the form filler, should be entered as stated on the form, or should be entered as the upper or lower bound of the range or some other option. If not already available, the researcher must provide for creation of field-specific data entry guidelines to support consistency and efficiency in data entry. Further, if not already available, the researcher must select, install, and test entry software. After a successful installation, the study-specific data entry screens can be created. Creation of study-specific data entry screens can also entail programming or otherwise setting up data storage structures, mapping from fields to data storage structures, and programming logic that runs during data entry including conditional sequence of fields on the entry screen, conditionally generated forms, and on-screen discrepancy checks. Afterward, the study-specific programming or configuration requires testing and documentation thereof. After the system is set up and validated, data entry is performed throughout the study conduct. Periodically, the data entry error rate is assessed and documented (Table 8.5).

TABLE 8.5

Tasks Associated with Key Entry Methods

Data Entry Tasks	All Cases	Special Cases
Specifying the data elements to be key entered	X	
Specifying and documenting the data entry procedures including field-specific guidelines	X	
Selecting entry software	X	
Installating data entry software	X	
Initial system validation (out-of-the-box testing)	X	
Designing the study-specific data collection form	X	
Mapping of the form fields to data storage structures		X
Programming or otherwise setting up data storage structures, data entry screens, and discrepancy checks		X
Testing and characterizing study-specific data entry system and procedures including documentation thereof	X	
Performing data entry tasks	X	
Ongoing assessment of data entry accuracy and documentation thereof	X	

The two tasks are marked as occurring in special cases, (1) Mapping of the form fields to data storage structures and (2) Programming or otherwise setting up data storage structures, data entry screens, and discrepancy checks. These tasks may be handled by the data entry system as part of designing the data entry screens - here the person setting up the screens has no control over the structure in which the data are stored.

Quality Assurance and Control for Data Entry

Anywhere there is a manual process (performed by humans), there is opportunity for error and inconsistency. For manual processes, prospective actions to assure quality are necessary but not sufficient to guarantee quality. To obtain consistent results, manual processes must be measured and monitored throughout the process (Zozus et al. 2015). In the case of data entry, the error rate must be measured and compared to limits appropriate for the study; feedback must be provided to entry operators and actions taken to improve the process where necessary.

Data entry processes are often structured so that entry operators make as few decisions as possible, reducing the task of entering data to keying data exactly as it appears on the form to limit the impact of human inconsistency. However, because of human inconsistency in filling out the form, there is opportunity for inconsistency with every field on the data collection form. To support consistency, data entry operators need instructions for handling instances where the data should or should not be entered as they appear on the form, and instances where it is impossible to enter data as it appears on the form. Common examples include the following:

- Handling a laboratory result written in a numeric field as >300. A numeric field would not accept the > sign. Furthermore, it may be important to know if the value were 302 or 350
- A laboratory result written on a form as 300–350 in a numeric field
- Correction of spelling; handling of abbreviations
- Conversion of units
- Handling date and time where time is recorded as 24:00
- Handling data written in the margins and text too long for the database field
- Handling instances where two responses are indicated in a *check one* field

It is important that manual operations be performed consistently; otherwise, bias may result. Thus, data entry operators, no matter how qualified, need data entry conventions that specify agreed ways to handle field-specific issues that come up during data entry.

Data entry conventions should be complete. After data entry, there should not be any instances of the entered data not matching the form that are not covered by the conventions. Any instances of entered data not matching the form should be explained by the data entry conventions. Thus, data entry conventions serve as both a quality assurance tool (a job aid) and a documentation of how data entry was done. Such conventions support traceability by documenting any changes to data during the data entry process.

Writing data entry conventions involves reviewing the data collection form, thinking of the likely issues, and documenting how they should be handled. Testing the form helps identify some of the common inconsistencies in form completion, but additional issues will often arise during the project. Thus, entry operators should be instructed to make the researcher or data manager aware of any issues that come up that are not covered by the guidelines.

Timing of Data Entry

It is much easier to track and check data once they are in electronic format. Further as described in Chapter 5, delaying data checking (often called data cleaning) decreases the achievable data quality. Data entry should occur as soon as possible after the data are collected. Once data are in electronic form, they are available for computer-based processing. Computer-based processing is consistent and not subject to errors to which manual processes fall victim. Thus, once data are in electronic form, manual processing should not be used if at all possible because manual processes are less consistent and more error prone.

Direct Electronic Capture of Data

As described in other sections of this chapter, manual steps are subject to inconsistency and error. The only true preventative strategy for these errors is to use measurement or observation techniques that provide direct electronic output, thus obviating the need for tasks to be performed by humans. If a measurement is made by an electronic device, it is best to obtain the data directly from the device through an interface or an electronic output file if at all possible. Devices designed for use in research often have mechanisms to get data electronically. One common mechanism involves devices that plug into a personal computer or use wireless methods to stream data out to software accompanying the device. Some of these software systems can be configured to write data directly to a database others write to durable media, whereas still others have various mechanisms for the researcher to obtain a file containing the data. Some devices only have an output display, for example, a dial, scale, digital display, or output to paper—this is not considered direct data acquisition from the device because the values from a paper printout generally must be manually entered. Similarly, copying the data from a display and entering it later involves manual steps; this transcription adds an opportunity for error.

Questionnaires completed electronically by respondents are another example of direct electronic acquisition. Although the respondents are manually entering data and subject to the challenges of manual data entry, respondent-completed forms are usually designed more carefully and simplify the entry process as much as possible. In short, developers of these systems pay much more attention to cognitive aspects of form and screen design and usability. These data are considered direct electronic acquisition, also called *eSource,* because the researcher obtains the data from the source electronically. Further, these systems offer the ability to identify and resolve discrepant data during entry (called *on-screen checks*).

Data that are obtained directly in electronic format are often not the only data collected for a study; thus, they must be associated with the correct experimental or observational unit, time point, or sample. The process of associating external data with other data for a study is called record linkage or more broadly data integration, and is described in Chapter 9.

From a workflow perspective, there are configuration or setup tasks as well as tasks associated with ongoing acquisition and integration of data from devices (Table 8.6). First, the researcher must specify the data elements to be captured directly in electronic form from devices and the data acquisition procedure. These two things inform the choice of direct capture methods, devices, and software. Once software is selected and purchased or otherwise obtained, if not already available to the researcher, the software must be installed and validated. Once the device or devices and software are tested, the study-specific data acquisition can be configured in the device and associated software. The researcher either designs or learns the format and content of the data obtained from the device and either specifies a data transfer file or configures a direct interface, both of which require mapping of the form or file fields to the study-specific data storage structures. Sometimes, this mapping is saved as a file and called a control file; the *control file* is used by the study database to read and import incoming data files. The control file is the instructions used by the data system to import data into the study database. Whether accomplished through the control file or through other means, the imported data elements are associated with other study data such as experimental or observational units, time points, or samples. These associations are referred to as integrating the imported data with other data for the study. In the study database, programming or other setting up of the data storage structures and discrepancy checks is required. Finally,

TABLE 8.6

Tasks Associated with Direct Electronic Capture of Data

Direct Electronic Capture Tasks	All Cases	Special Cases
Specifying the data elements to be captured directly in electronic form from devices	X	
Specifying and documenting the data acquisition procedure	X	
Selecting the direct capture method(s), device(s), and software	X	
Installing software		X
Initial system validation (out-of-the-box testing) of device(s) and software		X
Designing the study-specific data collection form		X
Defining (or understanding) the data acquisition file		X
Mapping the form or file fields to data storage structures	X	
Programming or otherwise setting up the mechanism to import data into the study database and integrate it with other data for the study	X	
Programming or otherwise setting up data storage structures and discrepancy checks	X	
Testing and characterizing study-specific data acquisition and procedures including documentation thereof	X	
Performing data acquisition-related tasks (file import, record linkage, reconciliation, data discrepancy checks, and archival of received data)	X	
Ongoing assessment of imported data and documentation thereof	X	

the data acquisition procedure from data capture using the device through to integration of data with the study database need to be tested. Once the system is tested, data acquisition-related tasks can be performed. These include receiving and importing data files, *data integration*, *reconciliation* of imported data with other data on the study, data discrepancy checks, and archival of received data. Throughout the study, imported data are monitored through the aforementioned checks. The tasks marked as occurring in special cases in Table 8.6 are those which depend on whether necessary software is already installed and tested and whether the researcher has the option to configure study—specific data collection structures.

Summary

This chapter covered multiple ways of getting data into electronic format. This basics of each method were covered as well as the tasks required to set up any required software or hardware, to configure systems for study-specific data acquisition, and to perform the ongoing data acquisition tasks.

Exercises

1. OCR and OMR are fully automated approaches to get data into electronic format. True or false?

   Correct answer: False, recognition-based methodologies often require a human verification step.

2. Which of the following methods reads handwritten characters?

   a. OMR

   b. OCR

   c. ICR

   d. None of the above

   Correct answer: (c) ICR.

3. Which of the following methods reads forms completed with checkboxes?

   a. OMR

   b. OCR

   c. ICR

   d. None of the above

   Correct answer: (a) OMR.

4. Which of the following methods reads machine-printed text?

   a. OMR

   b. OCR

   c. ICR

   d. None of the above

   Correct answer: (b) OCR.

5. A researcher wants to encode the form name and the study time point into a bar code label that is preprinted on the study forms. Which of the following is the best option?

   a. One-dimensional bar code only

   b. One-dimensional bar code with the information also printed on the form

   c. Two-dimensional bar code only

   d. Two-dimensional bar code with the information also printed on the form

   e. None of the above

   Correct answer: (b) One-dimensional barcode with the information also printed on the form.

The amount of information is small and can be encoded in a one-dimensional barcode, and using a redundant channel of communication by also printing the information on the form provides a human-readable version in case the barcode is damaged.

6. A researcher wants to use a smartphone application to dictate his or her field notes rather than keeping written records. Which of the following is he likely to need?

   a. A way to get the files off of his or her smart phone
   b. A way to preserve the files
   c. A way to label the files with the date and location of the observation
   d. A way to read the files on his or her computer
   e. All of the above

   Correct answer: (e) All of the above.

7. Rank the following in terms of achievable data quality from highest to lowest, assuming all computer systems have been tested and are working properly.

   a. Direct electronic acquisition
   b. Double data entry only
   c. Single entry with on-screen checks
   d. Single entry alone
      i. a, b, c, d
      ii. b, a, c, d
      iii. c, d, b, a
      iv. d, b, a, i

   Correct answer: (i) a, b, c, d. Direct electronic acquisition has the highest achievable data quality because the capture is the source. Of the manual entry processes, double data entry provides the highest quality, followed by single data entry with on-screen checks. Single entry alone is associated with the lowest quality of all of the methods.

8. In which of the following is the second entry operator blind to the values entered by the first entry operator?

   a. Double data entry with interactive verification
   b. Double data entry with independent verification
   c. Double data entry with third-party compares
   d. (a) and (b) only
   e. (b) and (c) only

   Correct answer: (e) (b) and (c) only. In interactive verification, the second entry operator sees the values entered by the first entry operator and makes the decision as to the correct response.

9. The sponsor of a study performs a site visit to the researcher's institution to see how the study is going. The sponsor voices concern that the data are only single entered and may be of low quality. What things might the researcher do to assure that the data are of an acceptable quality level?
   a. Review a random sample periodically and calculate an error rate
   b. Add on-screen checks to the data system
   c. Change to a double data entry process
   d. All of the above

   Correct answer: (d) All of the above. Any of the options (a) through (c) would help. If a random sampling if the error rate is too high, the researcher may need to move on to options (b) or (c).
10. Which of the following are necessary to explain changes to data during data entry?
   a. Standard operating procedure
   b. Data entry conventions
   c. An audit trail
   d. None of the above
   e. (b) and (c)

   Correct answer: (e) (b) and (c). Conventions explain what was done and the audit trail provides the date, the value before the change, and the user account from which the changes are made. SOPs are usually at to high of a level to explain data value-level changes.

References

Anderson P and Wortman B. (1998). Certified Quality Auditor (CQA) Primer. Quality Council of Indiana.

Nahm M. Clinical research data quality. In *Clinical Research Informatics, Health Informatics*, eds. R.L. Richesson and J.E. Andrews. London: Springer-Verlag, 2012, pp. 175–201.

Palmer RC. *The Bar Code Book: Fifth Edition—A Comprehensive Guide To Reading, Printing, Specifying, Evaluating, And Using Bar Code and Other Machine-Readable Symbols*. Bloomington, IN: Trafford Publishing, 2007.

White DJ, King AP, and Duncan SD. Voice recognition technology as a tool for behavioral research. *Behavior Research Methods, Instruments, & Computers* 2002; 34(1):1–5.

Zozus MN, Pieper C, Johnson CM, Johnson TR, Franklin A, Smith J, and Zhang J. Factors affecting accuracy of data abstracted from medical records. *PLoS One* 2015; 10(10):e0138649.

9

Data Structures

Introduction

Chapter 8 describes methods and considerations for getting data into electronic format. There are many different ways to structure and transport data once they are in electronic format. This chapter covers structures for data collection, exchange, and storage as well as two closely related topics, data integration, and processes around data exchange. This chapter prepares the reader to choose appropriate data collection formats and understand how they relate to common data storage or organization schemes in computer systems. Further, the reader will gain understanding of the basics of data integration and data exchange.

Topics

- Data collection formats
- Data storage structures
- Data integration
- Data exchange structures and concepts

Data Collection Formats

There are often different ways to collect the same data, for example, if a study collects the level of corrosion or other build up inside a pipe as none, mild, moderate, or severe, the response can be collected as a *fill-in-the-blank* or a *check only one* structure. As there are a limited number of mutually exclusive choices, a check only one structure is the most appropriate. Common data collection structures include the following:

Free text: Free text fields are often called write-in or fill-in-the-blank fields. Free text fields can be short for one or two word responses, or they can be long for large amounts of text. The latter when large enough for one or more paragraphs is sometimes referred to as a *narrative text field,* though the distinction is somewhat arbitrary because most systems allow the form creator to set any character limit.

Structured fill-in-the-blank: Structured fill-in-the-blank fields contain free text within the bounds of some structure. For example, date fields are often displayed on a paper form as __ __ /__ __ __ / __ __ __ __ (dd/mon/yyyy), where dd indicates that a two-digit day is expected, mon indicates that a three-character abbreviation for month is expected, and yyyy indicates that a four-digit year is expected. Several varieties of date formats are used in practice; the one shown decreases ambiguity by collecting the month in character rather than numerical format. Other examples include time collected as __ __: __ __ (hh: mm), where hh indicates a two-digit hour and mm indicates a two-digit minute. A 24-hour clock is often used to decrease morning versus evening ambiguity. Phone number is yet a third example of structured fill-in-the-blank and is sometimes as (__ __ __) __ __ __ - __ __ __ __. Fields where a numerical code is to be written as shown in the pathogen code example in Figure 9.1 are another example of structured fill-in-the-blank. Structured fill-in-the-blank can be enforced in electronic systems, for example, the acceptable responses for the month can be restricted to the digits from 1 to 12 or the three-letter abbreviations for months.

Check all that apply list: Check all that apply lists are a structure for capturing discrete data for which one or more option is expected from respondents. For example, the race question in Figure 9.2. The United States Office of Management and Budget has a standard for race data collection that dictates five high-level categories; respondents are given the opportunity to select multiple responses. On a data entry screen, a check all that apply list can be implemented as a list of checkboxes where marking multiple response items is permitted, as a yes/no question for each response if both affirmative and negative responses are needed, or a drop-down list that permits selection of multiple response items. Where a list of checkboxes allowing selection of more than one, the convention is to represent the response field (checkbox) as a square (question 4 in Figure 9.2). Response items for check all that apply lists should be at the same level of granularity unless more detail is purposefully of interest for some responses and less detail is desired for others.

Pathogen (complete only if pathogen isolated)

Type of Specimen	Method of Collection (check only one)	Results (see Pathogen Code List below)	Other* (if pathogen not associated with a code; specify Genus and Species)
1 Sputum:	☐1 Bronchoscopy ☐2 PSB ☐3 Mini-BAL	___ ___ ___ ___ ___ ___ ___ ___ ___ ___ ___ ___	
2 Transtracheal aspirates:	☐1 Bronchoscopy ☐2 PSB ☐3 Mini-BAL	___ ___ ___ ___ ___ ___ ___ ___ ___ ___ ___ ___	

Pathogen Code List (enter code(s) in Results column above):

Common:	101 *Staphylococcus aureus* (+)	**Pseudomonas aeruginosa:**	211 *Pseudomonas aeruginosa*
	105 *Streptococcus pneumoniae* (+)	**Legionella:**	221 *Legionella* (neutral)
	130 *Haemophilus influenzae* (-)	**Fungal:**	301 *Fungal*
Mixed anaerobic bacteria (aspiration) (±):	111 *Peptostreptococcus* spp. (+)	**Mycoplasma and Chlamydia:**	301 *Mycoplasma and Chlamydia*
	131 *Peptococcus* spp. (+)		
	132 *Prevotella* spp. (-)	**Mycobacterial:**	302 *Mycobacterial*
	134 *Bacteroides* spp. (-)	**Parasitic:**	303 *Parasitic*
	135 *Fusobacterium* spp. (-)	**Common—children:**	401 Respiratory syncytial virus
Uncommon:	136 *Acinetobacter var. anitratus* (-)		402 Parainfluenza virus types 1, 2, 3
	137 *Moraxella catarrhalis* (-)	**Common—adult:**	403 *Influenza A virus*
	103 *Streptococcus pyogenes* (+)		404 *Influenza B virus*
Enterobacteriaceae:	205 *Escherichia coli* (-)		405 Adenovirus
	206 *Klebsiella pneumoniae* (-)	**Other:***	199 Other (if pathogen is not associated with a code, enter 199 in Results column, and specify in Other* column above)
	204 *Enterobacter* spp. (-)		
	220 *Serratia* spp. (-)		

FIGURE 9.1
Example specimen and pathogen form.

1 Date of birth: ___ ___ / ___ ___ ___ / ___ ___ ___ ___
day month year

2 Sex:
○1 Male
○2 Female

3 Ethnicity *(check only one)*:
○1 Hispanic or Latino
○2 Not Hispanic or Latino

4 Race *(check all that apply)*:
☐ American Indian or Alaska Native
☐ Asian
☐ Black, of African heritage
☐ Native Hawaiian or other Pacific Islander
☐ White

FIGURE 9.2
Example demography form questions.

Check only one list: The ethnicity question in Figure 9.2 is an example of a check only one list. The list is logically constructed such that it is exhaustive and mutually exclusive. The response options in a check only one list are *mutually exclusive* and only one response is expected from respondents. Note that question 2 in Figure 9.2 violates this principle in that many would argue that two response options are not mutually exclusive for some individuals. In addition to mutual exclusivity, response items for a check only one list should also be *exhaustive* over the concept. Note that question 2 in Figure 9.2 again violates this principle in that for some individuals the two response options are not sufficient. Where only a subset of the possible responses are of interest and this subset might not apply to every respondent, an *other* option can be added to create exhaustiveness without allowing responses outside the subset of interest. Where response sets violate the principles of exhaustiveness and mutual exclusivity, the quality of the data are questionable. Similar to other list-type data collection structures, response items for check only one list should be at the same level of granularity unless more detail is purposefully of interest for some responses and less detail is desired for others. On a data entry screen, a check only one can be implemented as a *radio button* set, a drop-down list, or a set of checkboxes that limit selection to only one response option. The convention with a select only one list is to represent response fields as circles as was the case in car radios in the 1960s.

Visual analog scale: The visual analog scale (Figure 9.3) is designed to present the respondent a continuous rating scale. The data are collected as the measurement of the distance from an anchor to the respondent's mark. There is much discussion regarding the use of such a fine scale: can respondents really distinguish between 100 different categories? Converting a subjective measure into continuous data, may, in fact misrepresent the true data; however, they are used in practice. Implementing a visual analog scale in a data collection system will usually require custom programming because this is not a data collection structure that is supported by general-purpose commercial data collection systems. Systems designed to support electronic questionnaires completed by respondents are

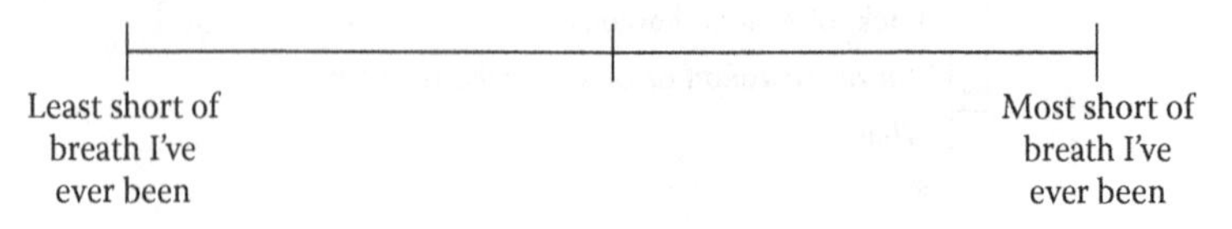
1. Mark the point on the scale that describes any shortness of breath you feel now.
Least short of breath I've ever been
Most short of breath I've ever been

FIGURE 9.3
Visual analog scale.

more likely to support this data collection structure. Where the data are collected directly from respondents electronically, this development is necessary because asking the respondents to enter a numerical measurement could invalidate the scale that was developed and tested as a VAS.

Image map: An image map collects spatial location data by allowing an observer to record where something is located relative to other items in proximity. Examples of image map data collection are used in collecting anatomical location, for example, the location of an injury, pain, lesion, or a procedure, in showing the configuration of vehicles after a traffic accident relative to each other and to landmarks, and the location of distinctive markings on objects. Image maps provide a user-friendly mechanism for the respondent to indicate the location or affected areas and avoid reducing the data. These diagrams can be spatially discretized into small areas that comprise the whole with each assigned a code; the data are entered according to which section(s) the markings fall. Consider a study conducted by a parks and recreation department on the impact of power boating on the local turtle population. The turtle population is sampled and the damage on the turtles' shells is recorded on an image map that shows the line-drawing layout of the scutes on the shell of resident species.

Image maps are easy to use with paper data collection forms but are more difficult to render on a computer screen; doing so requires custom programming in almost any system. Thus, the data are usually entered by the coded section according to some standard or the scheme devised for the study.

In Chapter 2, categories of data were covered. One category of data that is helpful in deciding a data collection structure is Stevens's scales. The data collection structure should match the scale of the data (Table 9.1). For example, collecting continuous interval or ratio data as discrete ordinal categories

TABLE 9.1

Data Type and Corresponding Data Collection Structure

Stevens's Scale Category	Corresponding Data Collection Structure
Nominal	Check only one list, check all that apply list, radio buttons, multiselect drop-down or select only one drop-down, type ahead for large pick lists, and structured free text for coded values
Ordinal	Check only one list, radio buttons, select only one drop-down
Interval	Structured free text, including integer and decimal numbers
Ratio	Structured free text, including integer and decimal numbers

reduces the information content and unless deliberate is not desirable. Such data reduction should be carefully considered because data reduction limits secondary uses for which the data might be repurposed after the study.

Data Storage Structures

The set operations research of David Childs, published in 1968, advanced the thinking that led to the concept of data independence, meaning that data could be stored separately from a software application, and the software application did not have to operate on the physical structure of the data (Childs 1968). Edgar F. Codd leveraged Childs' work on set-theoretic data structure to describe a relational data model in which data could be organized and stored independently from software applications in his seminal 1970 work describing the relational model (Codd 1970). Today, most software packages enable viewing or interacting with the data through such a logical data model. Data can also usually be imported and exported; the data can move independently of the software application.

This section covers a few simple principles from the relational model. Understanding these is important to ensuring that associations between data elements and between data elements and important context are not lost when data are entered and stored in a file, a spreadsheet, or a database. The relational model describes data as it would be structured in a table in terms of rows and columns. For any set of data elements, there are multiple ways that the data might be organized into tabular structures. Consider a study that is enrolling farms in a trial of two new varieties of corn compared to a standard variety. The farm eligibility data collected are shown in Figure 9.4.

Farm Eligibility

Summary of Inclusion and Exclusion Criteria

__ __
Farm ID

1. Were all of the inclusion criteria met?: ☐ yes ☐ no

2. Did any exclusion criteria apply to the farm?: ☐ yes ☐ no

3. If any criteria were not met, was a waiver approved by the study office?: ☐ yes ☐ no

If yes, please list the waiver approval date: __ __/__ __ __/__ __
day month year

FIGURE 9.4
Farm eligibility form.

TABLE 9.2

Farm Eligibility Data Structure Option 1

FARM_ID	Q1_ALL_ INC_MET	Q2_ALL_ EXC_MET	APPROVE_ REC	APPROVE_ DATE
32	1	0		
33	1	1	1	December 2, 2015

TABLE 9.3

Farm Eligibility Data Structure Option 2

FARM_ID	ELLIG_QUESTION	ELLIG_RESPONSE
32	1	1
32	2	0
32	3	
32	3-date	
33	1	1
33	2	1
33	3	1
33	3-date	December 2, 2016

Tables 9.2 and 9.3 show two different ways the data might be structured using data from two example farms. The first farm was given a study number 32; it met all inclusion and exclusion criteria. The second farm was given study number 33; it met all inclusion criteria but failed one exclusion criterion for which an exception was granted. Farm 33 was approved for enrolment on December 2, 2016.

Data structure option 1 has one row (also called record) per farm and one data element per column. Each data element on the farm eligibility form corresponds to a column in the table. The farm identifier (FARM_ID) is populated for each row in the table. The farm identifier is the mechanism through which all of the data are associated with a farm. In option 1, the farm identifier uniquely identifies each record in the table. This particular format makes comparisons between values on the form easier and makes it easier to tabulate descriptive statistics because comparisons are most easily done when data are on the same row. Further, because there is one data element per column, data quality checks are simpler to write.

Data structure option 2 has fewer columns and more rows. There are multiple rows per farm and one row per data element. There are four data elements on the farm eligibility form in Figure 9.4, and there are four rows per farm in the table. Each row in option 2 is uniquely identified by the combination of FARM_ID and ELLIG_QUESTION. The structure in option 2 more closely resembles the data collection form, so if data were entered into a spreadsheet structured in this way, data entry would likely be easier than option 1. However, software for data collection usually provides a feature to create data entry screens that look more like the form; these systems obviate the need to structure data tables to facilitate data entry. The option 2 format makes comparisons between values on the form more difficult because comparison between rows takes a bit more programming. Further, data quality checks take a bit more programming because valid value and data type checks only apply to certain rows in a column.

The ability to see a form or list of data elements and envision how the data may or does appear in a database table is important because relationships in the data are preserved through relationships within and between database tables. For example, in option 1 above, all data on a row are associated with the FARM_ID that appears on that row. Consider the farm characteristics in Figure 9.5. Seven data elements are collected, including the number of acres planted, indication of whether they are contiguous, zip code, number and date of soil samples, planned planting date, and study assigned farm identifier.

Figure 9.6 shows the farm irrigation log for study week 1. The irrigation log collects rainfall-equivalent inches of water from irrigation and irrigation-related comments for each day from planting (study day 1) until harvest. Thus, there is an irrigation log for each week of the study.

The soil samples referred to in Figure 9.5 are sent to a university-based agricultural lab for testing, and the researcher will receive a file with the soil analysis results. The file will contain the data from the lab's standard soil analyses report (Figure 9.7). Further, rainfall is a potential confounding variable, and the researcher will obtain monthly total inches of rainfall

Farm Characteristics

1. Plot randomization, acres and soil samples

Plot 1: **Seed #** ____ ___ **acres** **Number of soil samples:** __

Plot 2: **Seed #** ____ ___ **acres** **Number of soil samples:** __

Plot 3: **Seed #** ____ ___ **acres** **Number of soil samples:** __

2. Are acres planted for the study contiguous?: ☐ **yes** ☐ **no**

3. Zip code in which the majority of the acres fall: _____

4. Date soil samples taken: __/___/__ (**day** **month** **year**)

5. Planned planting date: __/___/__ (**day** **month** **year**)

6. Assigned farm ID for the study: __

FIGURE 9.5
Farm characteristics form.

Farm irrigation Log

Study week 1

__ __ Farm ID

Date	Inches (rainfall equivalent)	Comments
__/___/__ day month year	__	________
__/___/__ day month year	__	________
__/___/__ day month year	__	________
__/___/__ day month year	__	________
__/___/__ day month year	__	________
__/___/__ day month year	__	________
__/___/__ day month year	__	________

FIGURE 9.6
Farm irrigation log.

by zip code from the national weather service data archive. The researcher has also contracted a local pest control company to treat all study farms for the growing season and to survey the insect populations using an insect count by species on a monthly basis. Data files for the insect counts will be received by the researcher. The soil samples and soil analysis data are linked to the data from the data collection forms by the farm identifier. The rainfall data are linked to the data from the data collection forms by the zip code. If zip code was not collected, the weather service rainfall data could not be associated with each farm. Note, in reality, some zip codes contain very large areas and might not be useful for this purpose. If the soil samples were not labeled by the farm identifier, plot number, and sample collection date, the data could not be associated with the correct farm, and the yield data could not be analyzed with soil sample test results (Figure 9.8).

The relationships in the data are shown in Figure 9.9. There are many possible ways the data could be structured. The example in Figure 9.9 is not necessarily the best or the worst, it is but one of many options. The notation in Figure 9.9 is called crow's foot notation. The data elements listed in the boxes in the diagram represent columns in a table in a database. There is one box on the diagram per logical table in the database. Relationships between data in tables are shown in two ways. The first is by a line called an association. The second way that a relationship is shown is having the same column listed in two tables; this denotes that the two tables can be joined by matching up the values in the shared column. A crow's foot symbol on a box indicates that there are multiple rows in one table for each row in the table at the opposite end of the association. The exercises will more fully explore this example.

Phosphorus	12.78 ****
Potassium	357.56 *****
Iron	22.85 *****
Manganese	6.35 ****
Zinc	10.04 *****
Copper	9.18 *****
Boron	0.20 ***
Calcium	340.35 ***
Magnesium	363.25 *****
Sodium	374.10 *****
Sulfur	367.09 ***
Molybdenum	0.09 ***
Nickel	1.05 **
Aluminum	nd*
Arsenic	0.03 *
Barium	0.31 *
Cadmium	0.11 *
Chromium	0.05 *
Cobalt	0.14 *
Lead	3.78 **
Lithium	0.13 *
Mercury	nd*
Selenium	0.51 **
Silver	nd*
Strontium	1.47 *
Tin	nd*
Vanadium	1.15 **

Saturation Paste Extract			
pH	7.62****		
ECe (m-mho/cm)	2.42****		
Saturation Extract		millieq/L	
Calcium	161.5	8.1	
Magnesium	70.1	5.8	
Sodium	254.4	11.1	
Potassium	38.7	1.0	
cation sum		26	
Chloride	479	13.5	
Nitrate as N	35	2.5	
Phosphorus as P	0.7	0.0	
Sulfate as S	152.2	9.5	
anion sum		25.5	
Boron as B	161.5		
SAR	70.1		
Gypsum requirement (lbs/1000 sq ft)			101
Infiltration rate (inches/h)			2.99
Soil texture			
Sand	16.5%		
Silt	29.6%		
Clay	53.9%		
Lime (calcium carbonate)		yes	
Total nitrogen		0.092%	
Total carbon		1.161%	
Carbon:nitrogen ratio		12.6	
Organic matter (based on carbon)		2.32%	
Soil moisture content		21.7%	
Half saturation percentage		45.0%	

	Reference Ranges Extractable (mg/kg soil)		
Element	Low	Medium	High
Phosphorus	0–7	8–15	> 15
Potassium	0–60	60–120	121–180
Iron	0–4	4–10	> 10
Manganese	0–0.5	0.6–1	> 1
Zinc	0–1	1–1.5	> 1.5
Copper	0–0.2	0.3–0.5	> 0.5
Boron	0–0.2	0.3–0.5	> 1

FIGURE 9.7
Example soil analysis.

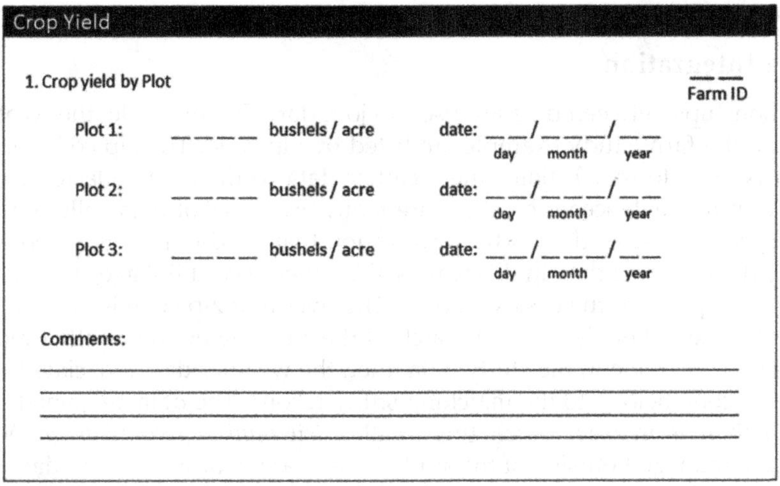

Crop Yield

1. Crop yield by Plot **__ __ Farm ID**

Plot 1: __ __ __ __ bushels / acre date: __ __ / __ __ __ / __ __ (day, month, year)

Plot 2: __ __ __ __ bushels / acre date: __ __ / __ __ __ / __ __ (day, month, year)

Plot 3: __ __ __ __ bushels / acre date: __ __ / __ __ __ / __ __ (day, month, year)

Comments:

FIGURE 9.8
Crop yield form.

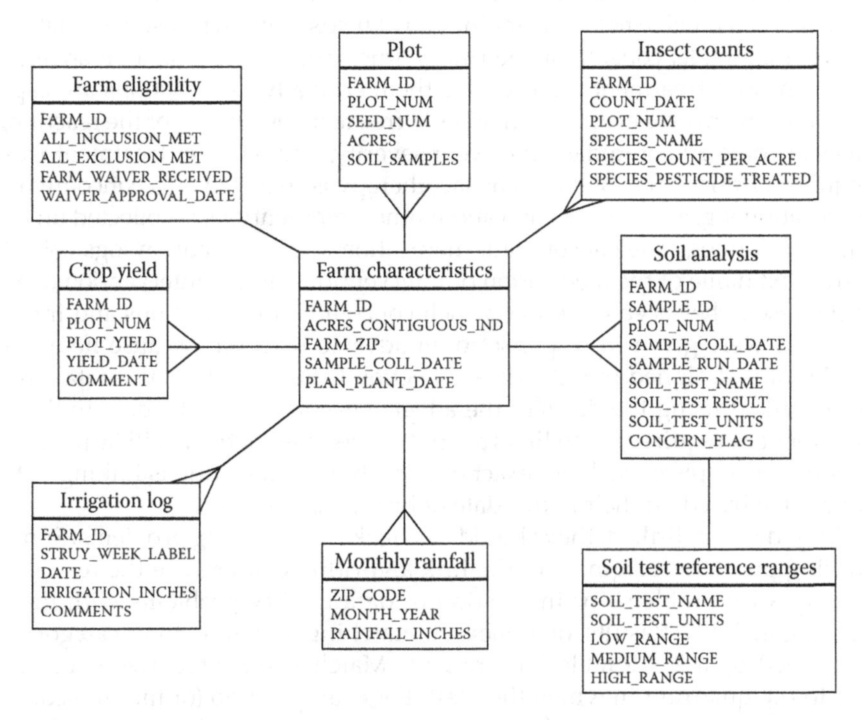

FIGURE 9.9
Example entity relationship (data structure) diagram.

Data Integration

Relationships between data are used to join data. For example, the weather data in the farm study example are listed by zip code. The zip code for the farm is used to match (join) the weather data to the data collected in the study. In research scenarios, there are many examples of data collected and managed outside of the study organization that need to be integrated with study data. Joining data in this way is also called record linkage, and the zip code example is of the easiest variety. The five-digit zip code is used to join records where there is an exact match. If the zip code on the weather record matches a zip code in the study data, then the weather data associated with that zip code belong to the matching study record. The data are joined only where there is an exact match; this is called *deterministic record linkage*. When data are managed outside of the study organization, or in different data systems, such links should be created so that the data can be joined. In many cases, researchers have the opportunity to design the exact link fields into a study; this is preferable where possible. When deterministic matching is used, records for which no match is found should be investigated for possible errors in the match field, or incompleteness in either of the joined data sets.

Record linkage is not always as easy as exact matching. In many cases, there is not a reliable field or combination of fields that can be used for matching. For example, consider a veterinary study that collects data on adverse reactions and treatment required for them. The adverse reactions are captured on an adverse events log that has a free text description of the reaction, the start date and stop date, the event severity, and whether treatment was required. Information about pharmacotherapy is collected on a concomitant medication log, and information about other interventions is collected on an intervention log. The concomitant medication and intervention logs collect a free text name of the medication or intervention and the intervention date. If the researcher needs to know which concomitant medications and interventions were given in response to an adverse reaction, because no association was designed into the data collection form or system, the data have to be linked using the dates of the adverse events where the data indicate treatment as required. Such links are not exact—associations will be missed. Another example of such nonexact or probabilistic matching is linking data for people based on their name, date of birth, and address.

After data are linked, they should be checked for linking problems. Some linking problems originate in the link algorithms; others are the result of data quality problems or interaction of data quality problems with linkage algorithms. Results of nonexact matching fall into four categories described by a 2 × 2 table (Figure 9.10). Matches are categorized according to the quadrant in which they fall. For example, two (or more) records matched by the algorithm that refer to the same real-world entity are classified as a true positive, whereas two (or more) records matched by the

		Match result	
		Matched by algorithm	Not matched by algorithm
Truth	True match	True positive (good match)	False negative (failure to match)
	True nonmatch	Fale positive (bad match)	True negative (good nonmatch)

FIGURE 9.10
Possible result of an attempt to match two records.

algorithm that do NOT refer to the same real-world entity are classified as a false positive. A match that the algorithm failed to identify is called a false negative, and a record that the algorithm correctly did not match is called a true negative (Figure 9.10).

The quality of the record linkage should be presented with research results because error-laden linkage can bias research results. In order to know if a match refers to the same real-world entity, ideally, the truth should be known. However, if a researcher had knowledge of the truth, record linkage would not be needed. One way to test the record linkage is to create a representative subset where the truth is known, test the match algorithm, and calculate false positive and false negative rates. The drawback of this approach is that the truth is not always accessible to the researcher, even for a subset. Further, a subset may not be representative, and assessing the matches for even a subset may be cost prohibitive. Other ways of testing include comparing match results from different algorithms and reviewing the records that link differently between the algorithms (this is a relative method), or selecting a random sample, and manually reviewing the links to see if they look reasonable (called the random sample method). The last two methods cannot provide the false positive and false negative rates because neither relative rules nor random samples make a comparison to known truth or to some approximation such as a gold standard.

Similar to the optical data entry methods described in Chapter 8, record linkage processes can be designed with a manual review step. Some record linkage software allows computing a confidence in the match and associating the confidence with the matched records. Thresholds can be set where the confidence is lower and these matches can be reviewed in hopes of decreasing false positives and false negatives. Likewise, computational methods have been developed to detect records that may not be good matches; these matches can also be flagged for manual review in the hope of decreasing false positives. Researchers needing record linkage beyond exact deterministic matching are referred to the Entity Resolution literature, for example, *Entity Resolution and Information Quality* by John Talburt (2011).

Data Exchange Structures and Concepts

Data require integration as described in the section "Data Integration" when they are collected or managed by different people, systems, or organizations. A precursor to the actual linkage and resulting integration is the transfer of data from one person, system, or organization to another. This section covers the case where this happens during the research project. Data transfer in the context of data sharing and secondary use is covered in Chapter 16.

Recall the discussion in Chapter 5 about data moving and changing: Where data are transferred from one person, system, or organization to another during the course of a study, the data are moving as described in Figure 5.5. The moving of data in this way is called a *data transfer*. There are multiple modes of data transfer. These include direct system interfaces, standards-based messaging of data or documents, and custom transfers of data. *Direct system interfaces* are often described as two systems that *talk* to each other. An example of such an interface would be where data collected with a device show up automatically in a research database without the sender or receiver having to do anything. Many such systems come with configurable interfaces. Where configurable interfaces are not available in a software package, development of direct interfaces requires some work initially to set up and test.

Use of the second modality, messaging can mimic a direct interface. Many systems support data exchange standards. For example, information systems in health care support *Health Level Seven* (www.hl7.org) exchange standards. These systems can be configured to recognize when data need to be sent to another system and can do so automatically. The messaging mode of data transfer tends to be used for small amounts of data transferred near real time. Transaction processing is similar to messaging, but requires sending all information necessary to process a transaction, usually a larger amount of information. For example, a message for a bank withdraw would not be actionable without all of the information; thus, such information transfer occurs by transaction rather than messaging. Sending lab results from a recent sample to the health record of the patient is a good example of messaging because lab results can be (and usually are) sent as they are ready, rather than waiting for results from all ordered tests. Messaging also works with documents. To use another example from health care, after a patient is seen by a specialist, the clinic note from the visit can be sent to the patient's primary care provider as a *structured document*. A document is a collection of data that must remain together because they are approved or otherwise authenticated together. Data standards also exist to support exchange of structured documents. Use of data standards where they exist can save time in exchanging data. Where direct interfaces are not possible and standards do not exist or are not yet supported by software, researchers must design a way to get data from one person, system, or organization to another.

Data transfers are simplified when two people or organizations are using the same software; many information systems have import and export functionality. This functionality creates an export file that can be read and imported by another installation of the software. Thus, the researcher does not have to write computer programs to write data out of one system and read them into another system.

When all of the above have been investigated and found not to exist, the researcher gains yet another data processing task. Moving data from one place to another requires locating the data to be sent, writing data out of the sending system into a format suitable for transfer, sending the data, locating or otherwise receiving the sent data, reading the received data, and placing them in the correct format and location in another system (Table 9.4). Of course, all of these steps need to be checked to assure that the desired data were sent, received, and integrated correctly. It is very much like packing a child up for their first solo train ride to visit a relative. There is checking to make sure that everything was packed, and checking to be sure of the safe arrival (Tables 9.5 and 9.6). Tables 9.5 and 9.6 list steps that commonly occur in data storage, data transfer and data integration.

As mentioned earlier, data can be sent as one message, document, or transaction at a time, for example, the results from running one sample, or from one observation. Data can also be sent in batches; a lab may send the data from all samples run in the prior month is an example of a batch data transfer. Decisions about how much data and how often data should be sent are parts of setting up data transfers. In the absence of data systems that support such tasks, it takes more effort to set up and manage real-time or very frequent data transfers than less frequent ones. The need for frequent transfers is usually balanced against the effort required. The rule of thumb is

TABLE 9.4

Steps to Transfer Data

	Data Transfer Task
Step 1	Locate the data to be transferred.
Step 2	Write the data values out to a data transfer file.
Step 3	Send the file from the source system to the destination system.
Step 4	Locate or receive the file.
Step 5	Read the file.
Step 6	Write the data values to the correct location and format in the destination system.

TABLE 9.5

Data Storage Tasks

Data Storage Tasks	All Cases	Special Cases
Annotation of data collection forms or other method of specifying table or file structure	X	
Specification of entity and referential integrity	X	
Specification of stored procedures		X
Write or configure DDL to create data tables (if using a database management system)	X	
Test data storage	X	
Acquire and load content for look up tables or knowledge tables used for coding		X

TABLE 9.6

Data Transfer and Integration Tasks

Data Integration Tasks	All Cases	Special Cases
Write data transfer specifications (file content and format).	X	
Decide and create mechanism for transfer if data, for example, shared cloud space or file transfer protocol site.	X	
Test file transfer.	X	
Specify, program, and test preload checks and exception reports for received data.	X	
Specify, program, and test programs or configuration that maps, reads, and writes data from transfer files to data storage location.	X	
Ongoing—retrieve or otherwise receive data and archive copy of data as received.	X	
Ongoing—execute preload data processing and work through exceptions with sender.	X	
Ongoing—load data.	X	
Specify, program, and test record linkage.	X	
Specify, program, and test postlinkage checks/data reconciliation and associated exception reports.	X	
Ongoing—run record linkage, reconciliation, and work through exceptions with sender.	X	

similar to that used to decide how often to check interrater reliability. Check at intervals no longer than that for which data loss is bearable and no shorter than that at which data can feasibly be checked. The decision about data transfer frequency is a compromise between effort or cost and risk mitigation.

When setting up data transfers, a decision also has to be made about how changes to data will be handled (Figure 5.5). If the source system will be making and documenting changes to data, the receiver may want only new or changed data or instead may want a complete copy of all of the data. The former is referred to as an *incremental transfer*, whereas the latter is referred to as a *cumulative transfer*. In addition, the receiver may require a copy of all changes to the data and information about those changes, also called the *audit trail*.

A custom data transfer is documented by data transfer specifications. The specifications describe what data are transferred, the definition, format, and units of each data element included, where the data are located in the data transfer file, and how data values and records are separated in the file. There are also specifications for the computer programs that write the data out of the sending system and specifications for *loading* the data into the receiving system. These specifications may change over time; thus, the multiple versions must usually be maintained. Transferred data should retain their association with the version of the specifications under which they were transferred. Often data are reformatted or otherwise operated on in transfers; when this occurs, association with the transfer specifications is required for traceability.

Additional practices for transferring data include checking the data. When data are written out of the sending system, the sender often prepares a simple report of the number of data elements and the number of records included in the transfer file(s). This report serves as a *packing list* to let the receiver know the amount of data to expect in the file. When the receiver obtains the data file(s), *preload data checks* are also performed. Received data files are archived, and received data are often read into a temporary or staging area for these checks. The initial checks include those for things that would prevent the receiver from reading, loading, or using the received data. For example, the initial checks include confirming that the number of data elements and records on the *packing list* was received; failing this check might signify that the wrong file or files were sent. Other checks might include confirming that the data values are in the specified formats, and that valid values were received; failing this check may signify a problem reading the data or misunderstanding the specifications. Problems at this stage usually require new files to be sent; thus, use of the temporary staging area for these checks. Once received data have passed these checks, they are usually integrated and loaded into the data system.

Managing data for a study where data are received from different places brings with it additional tasks for maintaining research reproducibility. For example, tracking versions of specifications, archival of transferred data files, and audit trails from source systems are required to maintain traceability. The process described here of writing data out of one system and into another system is called *extract transform and load* (ETL). Although the ETL topic belongs to the data processing chapter (Chapter 10), it is covered here because the data often change the structure from the source system to the transfer mechanism, and again when they are written to the receiving system. Maintaining traceability and associations in the data necessitates understanding how the data move through the data structures chosen for transfer, and in fact, maintaining traceability and associations through the ETL process.

Summary

This chapter covered data collection formats, data storage structures, data integration, and data exchange structures and concepts. It covered data structures primarily from a relational data model perspective. Three themes are central to this chapter. The first theme is that the researcher, along with identifying data to be collected, must also identify any necessary associations between data elements, that is, how the data need to be connected. The second theme is the importance of being able to look at a data collection form and visualize how the data might appear in a tabular form and vice versa. The ability to see and think in data structures helps the researcher assure that important associations are maintained. The third theme is that data transfers should be specified in advance, orderly, and well documented.

Exercises

1. Consider the form below. Which of the following is the best data collection structure for the case justification data element?

   a. Check all that apply checkboxes

   b. Check only one checkboxes

   c. Check only one drop-down list

   d. Check all that apply drop-down list

   Correct answer: (b) Check only one checkboxes. The two responses on the form are mutually exclusive; thus, a check only one structure should be used. The check only one checkboxes take one less click than the check only one drop-down list, so they are preferable.

2. For the form in question 1, which of the following is the best data collection structure for the clutter image rating data element?

   a. Free text

   b. Semistructured free text

   c. Radio buttons

   d. Table or log

   Correct answer: (c) Radio buttons. The options on the form are mutually exclusive; thus, a check only one structure is the most appropriate. Radio buttons most closely match the format on the data collection form.

3. How could the date fields on the form be improved?

   Correct answer: Date is a semistructured data collection field. The field on the form could be improved by having a separate space for each character, for example, ___ / ____ / ______ rather than only the slashes to divide day, month, and year. Further, the field could be improved by labeling the blanks, for example, dd/mm/yy, and by using a nonambiguous structure, for example, dd/mon/yyyy.

CUYAHOGA COUNTY Board of Health
PREVENT · PROMOTE · PROVIDE

Cuyahoga County Hoarding Connection
Data Collection Form
Vince Caraffi, RS, MPH
Cuyahoga County Board of Health
Tel: 216-201-2001 x1209
Fax: 216-676-1317

Case Information

Date of Visit: ___/___/___

For Offical Use Only
Case Status: ☐ Open ☐ Closed ☐ Denied

Case Justification
☐ Justified Report ☐ Un-Justified Report

Hoarding Type
☐ Animal ☐ Material ☐ Both

Lead Agency: ____________ Lead Contact: ____________

Personal Information

First Name: ____________ Middle Initial: ____ Last Name: ____________

Date of Birth: ___/___/___ Sex: M F

Ethnicity
☐ Hispanic ☐ Non-Hispanic

Race
☐ Caucasian ☐ African American ☐ Asian ☐ Other

Primary Contact: ____________ Secondary Contact: ____________

Is this person the Primary Hoarder?
☐ Yes ☐ No

Is this person (Spouse/Significant Other) a Hoarder?
☐ Yes ☐ No

Address Information

Street Number: ______ Street Name: ____________

Apt #/Suite: ______

City: ____________ Zip Code: ______

Clutter Image Rating

Survey Date: ___/___/___ Survey By: ____________

CIR Living Room: 1 2 3 4 5 6 7 8 9
CIR Bedroom: 1 2 3 4 5 6 7 8 9
CIR Kitchen: 1 2 3 4 5 6 7 8 9
CIR Bathroom: 1 2 3 4 5 6 7 8 9

Intervention

Intervention Date: ___/___/___ Intervention By: ____________

Agency: ____________

Intervention Type
☐ Clinical ☐ Enforcement ☐ Other

(Rev. 10/11)

4. Identify the cognitive principle(s) used in question 3?

   Correct answer: Proximity–compatibility principle would require that the blanks be labeled on the form, as would the principle of distribution of information across internal and external representations. Further, printing the directions on the form (structure of the blanks and the labels for the blanks) reduces the cognitive load of the form filler.

5. Given the hoarding data collection form from question 1, if the process was to complete one form per reported hoarder, and the personal information data elements were stored in one table with one data element per column, which two responses indicate that many records are possible in the table?

   a. One record per hoarder
   b. Multiple records per hoarder
   c. One record per report
   d. Multiple records per report

   Correct answer: (b) and (c) only. If there is only one report for a hoarder, there would be one record per hoarder. However, because multiple reports are possible for a hoarder, there may be multiple records per hoarder and one record per report.

6. There are multiple ways that the cluster image rating data may be stored in tabular form. If one of the columns in such a table indicated which room the rating pertains to and another column stores the numerical rating, how many columns are in the table if HOARDER_ID and SURVEY_DATE are included for referential integrity?

   Correct answer: Five. The columns are HOARDER_ID, SURVEY_DATE, the room rated, the CIR rating for the room, and the individual completing the survey.

7. For the example given in question 6, how many rows would there be in the cluster image rating table for each report?

   Correct answer: Four, one row for each room rated.

8. A social services researcher is interested in expanding the hoarding data collection project to 10 additional counties in each of five states and adding 6-month and 1-year follow-up visits to assess the longer term impact of interventions. In each state, one of the county mental health services programs will collect the form and enter the data into their local case management system. Is record linkage required?

   Correct answer: No. Record linkage would only be required in this case if the researcher expected that individuals from one county would relocate to another county in the study and self-report or be reported in the second county.

9. In the study described in question 8, would data transfers be required?

   Correct answer: Yes. The researcher would need to receive data from each state.

10. The researcher in questions 8 and 9 is also interested in obtaining the Society for the Prevention of Cruelty to Animals (SPCA) data for animal hoarding cases. He or she is interested in the percentage of animal hoarding cases reported that involved SPCA resources, and for those that did, the number and disposition of the involved animals. Is record linkage required to do this?

    Correct answer: Yes. To match SPCA data to a particular hoarder.

11. Given the research scenario in question 10, what data elements would be needed for record linkage?

    Correct answer: The record linkage would likely use name and address information.

12. Given the research scenario in question 10, would data transfer be needed? If so, describe any considerations.

    Correct answer: Yes. Data transfer from the SPCA to the researcher would be required. As SPCA investigations and rehabilitation and rehoming of surrendered or confiscated animals take time, the data would likely change over the course of the study. Thus, decisions would need to be made about incremental versus cumulative transfers, and about transfer frequency.

References

Childs DL. Description of a set-theoretic data structure. University of Michigan, Technical Report 3, 1968a, Ann Arbor, MI.

Childs DL. Feasibility of a set-theoretic data structure. University of Michigan, Technical Report 6, 1968b, Ann Arbor, MI.

Codd EF. A relational model of data for large shared databanks. *Communications of the ACM* 1970; 13(6):377–387.

Talburt JR. *Entity Resolution and Information Quality*. Burlington, MA: Morgan Kauffmann, 2011.

10

Data Processing

Introduction

Chapter 8 notes that for all but the smallest of data sets, a computer is needed for data processing, storage, and analysis. This is particularly true when the data volume is high or significant processing is needed. This chapter presents manual and computerized methods for *data processing*—those operations performed on data to prepare the data for analysis. Data processing operations include cleaning, imputation, transformation, coding, standardization, integration (covered in Chapter 9) and enhancement. Data processing creates special traceability concerns because although a data processing operation for a particular data element looks at all data values for that data element, only values meeting the criteria for the operation will be altered. Maintaining traceability requires tracking data changes at the data value level. Chapter 10 prepares the reader to choose and apply appropriate data-processing methods and to prospectively identify error sources and appropriate prevention or mitigation strategies.

Topics

- Data cleaning
- Imputations
- Data standardization
- Mapping
- Formatting
- Conversions and calculations
- Data enhancement
- Data processing and research reproducibility

Data Cleaning

Data cleaning is the process of identifying and resolving discrepant data values. In most cases, data errors—data values that inaccurately represent truth—are difficult to detect because we have no way of knowing truth. In practice, discrepant data values are identified because they are easier to detect. A *discrepant data value* may or may not be an actual error. Once a discrepant data value is identified, the source of the data is sometimes consulted in attempts to resolve the discrepancy. Discrepancies that result in a change to the data imply that the discrepancy was indeed an error. Other discrepancies, however, may be confirmed as a correct value and do not result in a change to the data. The term error applies only in the former case.

Chapter 7 briefly discussed data accuracy and mentioned that surrogates for accuracy are often used in practice. One such surrogate is the use of data discrepancies as an indicator of the potential number of errors. Data cleaning uses the same principles but does so not in a summative data accuracy measurement context but in a formative find-and-fix or find-and-notify context.

Data cleaning is a very common data processing task. There are several variations of data cleaning used in practice. All of them involve identification of data discrepancies. The variation comes from what is done after a discrepancy is identified. The first mode of data cleaning flags the discrepancies in the data, so that data discrepancy metrics can be aggregated, reported, and potentially dealt with in the analysis, that is, as part of data quality assessment. In this case, called *detect and flag* mode, a resolution to the discrepancy is not pursued. A second variation is called *detect and prevent.* In this variation, detected discrepancies may or may not be flagged in the database, but they are reported to the source of the data with the expectation that the recipient makes efforts to prevent or identify and correct the discrepancies in future data—no changes are expected for data that have already been collected and received. The third variation is called *detect and resolve.* Detect and resolve data cleaning involves notifying the source of each discrepancy with the expectation that each discrepancy is responded to either correcting or confirming the data value. Depending on agreements regarding who makes and tracks changes to data, the corrections may be implemented by the source of the data sending corrected data values or by the recipient using the response to the discrepancy notification as the documentation supporting the data change.

As described in Chapter 8, some discrepancy identification can take place during data entry as a real-time process that runs as data are entered, so that discrepancies can be identified and corrected immediately or later during double data entry or verification processes. The closer to the collection of the data the cleaning occurs, the better (Chapter 5). When data are cleaned later in time from the observation, asking, measuring, and recording the value, the opportunity to correct data diminishes, and the cost of correcting data values increases. Prior to widespread collection and entry of data via web-based systems, data were recorded on paper forms and sent to a central data center for entry. In this case, discrepancies were detected after data entry and much later in time than the data collection. When research studies were used detect and resolve processes, each detected data discrepancy was listed and communicated to the source of the data, and a resolution was expected. Often, when the resolutions were entered into the central database, additional discrepancies were created or subsequently identified. Tracking each discrepancy through this process was time intensive and expensive.

Discrepancy detection may be best described as an exercise in opportunistic creativity. Discrepancies are detected through some comparison: comparison to a source of truth, comparison to some standard or expectation, comparison to an independent source of the same information, comparison to an upstream version of the same information, and comparison to other values in the same study with which they are expected to exist in some relationship. The opportunistic creativity comes in identifying relationships and sources that can be exploited for such comparisons. A special case of comparison exists. When the source for comparison is truth or when the discrepancies identified by such a comparison are incompatible with reality, the identified discrepancies are errors rather than just potential errors. The closer to truth the source of comparison is, the closer to accuracy is the assessment. The more independent the source for comparison is, the more likely that the set of discrepancies identified is complete (Table 10.1).

Comparison to a source of truth is a comparison to something external to the study data. This comparison requires a recording of the event of interest as described in Chapter 5. The bar is high for what we call a source of truth. In the case of a survey about past exposure, asking the respondent a second time would not count, nor would using a second observer. The only source of

TABLE 10.1

Comparisons Used to Identify Data Discrepancies

Data Values Are Compared To	Strength of Comparison	Logic
A source of truth	External comparison	Do the values match?
Some known standard or expectation	External comparison	Do the values match?
An independent source of the same information	External comparison	Do the values match?
An upstream version of the same information	Internal comparison	Do the values match?
Other values in the same study with which they are expected to exist in some relationship	Internal comparison	Dependent on the relationship between the data elements being matched

truth would be a recording of the event of interest. This comparison is rarely done because such a source of truth rarely exists. However, prior to settling for lesser comparisons, the lack of such a source should be verified.

Comparison to some standard or expectation, also an external comparison, is the second strongest comparison because this comparison can identify errors and, in doing so, can measure accuracy. For example, human body temperatures of 120°F and above are incompatible with life. Thus, a recorded temperature of 120°F or higher in a person represented as alive is with almost certainty an error. Such standards are not available for every data element in a data set; thus, this accuracy assessment only applies to the data elements for which the standards exist. Further, this comparison will not detect all errors—data values in error and within the expected range will not be detected. Thus, this comparison yields a partial count of data errors. Even so, such standards should be sought and used when they are available for data elements.

Comparison to an independent source of the same information is a commonly used external comparison when an independent source is available. An example of such an independent source is comparing data from two independent observers of an event as done in assessing interrater reliability. In such a comparison, when the two values do not match, a discrepancy is noted; such a discrepancy can be compared to a source if one exists to determine the correct value, but without a source, nothing can be said about accuracy except that at least one of the data values is in error.

Comparison to an upstream version of the same information is an internal comparison, in which the source for comparison is an earlier version of the same information. This comparison is often used to measure the fidelity of data processing steps. For example, comparing the original data collection forms with entered data will detect data entry errors and support their correction. Such a comparison is only an assessment of the accuracy of a part of the data stream and will leave errors in the data undetected. In the above example, data recorded incorrectly on the form would not be detected.

Comparison to other values, in the same study with which they are expected to exist in some relationship, is an internal comparison and can be performed on any data set. An example of such a comparison would be that telemetry data on an experimental unit cannot be date stamped earlier than the installation date of the telemetry device. To perform the comparison, rules are written similar to comparison with known standards. For example in the farm study in the previous chapter, a rule that identifies instances of a farm reporting a zip code that is not in the reported state would be a likely rule. For any study, hundreds of such rules can usually be identified. There are, however, several weaknesses in this approach. The first is that these rules will not detect errors that happen to comply with the rule, for example, a wrong zip code that is actually in the reported state. The second is that these rules in this category do not identify instances of error, but only identify potential errors. Thus, comparison to other values in the same study is not a measure of accuracy; such a comparison is only a surrogate for accuracy.

Data quality has associated costs including both (1) the cost to achieve a desired level of quality, that is, error prevention costs and data cleaning costs and (2) the cost incurred for failing to achieve a necessary level of quality. The cost of quality (ASQ 1999) ideas originated and flourished in manufacturing through the work of thought leaders such as W. Edwards Deming and Joseph M. Juran, and have since been applied to other areas, for example, accounting (Spiceland et al. 2010) and software development (Boehm and Papaccio 1988), where it has been shown that correction costs increase exponentially the further downstream an error is detected and corrected (Ross 2012, Walker 2012). Walker provides an example of these costs with address data in the context of a company that ships products using consumer supplied mailing addresses (Walker 2012). Briefly, the 1–10–100 rule, a common rule of thumb in data quality circles, conveys that there is an order-of-magnitude increase in cost as one goes from the cost to prevent an error, to the cost of finding and fixing the same data error after occurrence, to the cost of a failure due to the error. In fact, in the author's experience, an *on-screen* error check costs a few dollars to implement and address during data entry, and costs an estimated US$35 if the data discrepancy is identified after data have been submitted to the data center (Kush et al. 2003), and further costs much more if caught during or after analysis. In the context of significant cost pressure in research, the likelihood of a serious data error and the potential impact should be weighed against the cost of preventing or fixing such data errors. Especially when weaker and more costly types of data cleaning may be avoided all together by designing upstream data cleaning into the design of a study (Chapter 5).

Data cleaning can also be performed manually, often called *manual review*. Examples include but are not limited to a human reviewing data listings in effort to identify odd-looking data or a human systematically executing prespecified rules to identify missing and inconsistent data. The former does not count as a comparison like those listed in Table 10.1 because a source for comparison does not exist; rather, it is a matter of opinion. The latter case of a human executing prespecified rules is a comparison; however, using a human in this way is costly and error prone. Even in the presence of these problems, for small studies, human manual review may be the best choice. In such a case, systematizing manual review tasks (quality assurance) and monitoring the error rate of the review (quality control) can mitigate errors.

Obvious corrections are changes to data that are made without a resolution or confirmation process. Obvious corrections can be made by man or machine. When made by humans, they are subject to the considerations above. When made by machine, the risk of human error is decreased. However, all changes made without confirmation have some risk of being incorrect. For traceability, obvious corrections should be documented.

As a form of data processing, data cleaning may add values to a database, for example, flags indicating that a value has failed a data quality check. Through the resolution of identified errors or discrepancies, data cleaning may also change data values. Both of these operations performed on the data result in new or changed data and must be tracked, so that data values can be traced back to the original value with all changes recorded. Further, any tasks performed by humans are associated with an error rate, and any systematized changing of data has opportunity for bias. Therefore, a statistician familiar with the planned analysis should review all such plans.

From a workflow perspective, data cleaning involves setup tasks, for example identifying sources of comparison and programming the comparison, tasks involved with the actual handling of the discrepancies, and tracking or reporting the discrepancy disposition. The setup tasks in data cleaning start with specifying the data elements to be checked for discrepancies and deciding for which if any discrepancies resolutions will be sought, that is, which mode of data cleaning will be used. As part of identifying the data elements to be cleaned, the comparisons to be performed are also determined, and the data cleaning procedure is specified and documented. The comparisons are then programmed or otherwise configured, then tested. When the number of discrepancies is expected to be large, for example, more than a few hundred, formal tracking will likely be needed. Such tracking and reporting of discrepancy disposition will likely require programming, or other configuration, or setup and testing. For large studies with a significant amount of data cleaning, tracking and reporting are necessary to manage the process of assuring that all discrepancies to be communicated to others are communicated, that resolutions are received, and that necessary database updates are performed. For such tracking, clear articulation of the discrepancy dispositions is necessary. For example, a discrepancy may have the following statuses: new, communicated, confirmed as is, or data changed. These statuses are mutually exclusive so that a discrepancy can have only one status at any point in time, such that a report of the statuses reflects the work completed and the work to be done. Discrepancies at each status can be listed individually to create work lists. When large amounts of data are managed and cleaned, discrepancy tracking is often used in this way to manage the data cleaning process, that is, performance of the data cleaning tasks. Because the discrepancy resolutions are often applied by humans, ongoing assessment of accuracy is necessary. In addition, fully testing every comparison is difficult because any comparison may have several logic conditions; testing multiple condition rules requires multiple test cases, and in the context of a research study, it is rare that all test cases are created and used. Thus, where programmed discrepancy checks are used, the comparisons are often monitored by a report that list how often each comparison identified a discrepancy. Those identifying a high number of discrepancies or no discrepancies are investigated to make sure that the comparison is working correctly (Table 10.2).

TABLE 10.2

Tasks Associated with Data Cleaning

Data Cleaning Tasks	All Cases	Special Cases
Specifying the data elements to be checked for discrepancies and of those for which if any will resolutions be sought	X	
Specifying and documenting the data cleaning procedure	X	
Specifying the comparisons to be performed	X	
Programming or otherwise configuring the comparisons	X	
Testing the comparison logic	X	
Programming tracking and reporting of discrepancy disposition		X
Testing tracking and reporting of discrepancy disposition		X
Performing data cleaning tasks	X	
Ongoing assessment of data cleaning accuracy and documentation thereof	X	

Imputations

Imputation is the process of systematically replacing missing values with an estimated value. For example, one of the older methods for missing data is to carry the last observation forward, called Last Observation Carried Forward (LOCF). A method commonly used for missing months in dates is to consistently assume January or June, and similarly, missing day values are sometimes imputed as the first or fifteenth of the month. There are many methods for imputation and whole books devoted to the topic. Imputations have the potential to bias data; thus, imputation should be performed systematically and guided by a statistician familiar with the planned analysis. For this reason, imputation is often done after any efforts to clean data are complete and just before analysis. Like data cleaning, imputations change data values and must be tracked to support traceability.

Standardization

Data standardization in information systems parlance is a process in which data values for a data element are transformed to a consistent representation. For example, consider a data element that collected the state in which a research facility was located. If the set of values contained the following: Ca, CA, California, Calif., NY, New York, NewYirk, and so on. The data would be more easily used if the values were associated with standard codes for the states, such as a two-letter code CA, and NY, or the full name California and New York. The process of applying standard codes to discrete data is called *coding*. Coding, also called mapping (see the section "Mapping"), is one type of data standardization. It can be done manually for very small amounts of data. If done manually, coding is subject to the inconsistency and human error associated with tasks performed by humans and thus requires both quality assurance and quality control. Coding may also be accomplished algorithmically by computer programs that parse the text and look for matches. The algorithms are developed and tested for specific types of data, for example, names of locations, diagnoses, or medications, and can achieve match rates in the 80%–90% range or above. The nonmatches are sometimes handled by a human and in other cases left uncoded.

Semistructured fields can also be standardized. For example, telephone numbers are often entered in multiple formats, for example, (xxx) xxx-xxxx, xxx.xxx.xxxx, xxx.xxx.xxx, and so on. The pattern variations like the lexical variations in the examples above are different representations, for example, (xxx) xxx-xxxx versus xxx.xxx.xxxx, or they can be caused by errors, for example, xxx.xxx.xxx. Continuous data may also require standardization. For example, data values for a data element may appear at different levels of precision. This may be the legitimate result of data from instruments with different levels of precision, or it may be the result of inconsistency in data recording. The former does not need to be standardized; doing so would reduce the information contained in the higher precision data values. The latter should be standardized to the actual level of precision of the instrument.

Standardization like imputation and cleaning alters data values and requires a systematic approach to avoid bias and tracking changes to data values to support research reproducibility.

Mapping

Mapping in information systems parlance is the process of associating data values to another set of data values, usually but not always using standard codes. Some types of mapping are reversible meaning that they do not result in information loss. Others result in information loss; depending on the uses to which the data are put, information reduction may be acceptable or even desired. The categorization of different types of mapping comes from mathematics where a function is defined as a way of mapping each member of one set onto exactly one member of another set. Mapping also draws from the set theory branch of mathematics that defines and describes operations on sets as units into themselves without regard to the nature of their members. Members are the elements in a set, for example, in a set containing the colors red, blue, and green, there are three members, *red, blue,* and *green*. The phrase *each element* in the definition of a function means that every element in the first set is related to some element in the second set. If a member of the first set was not matched to some member in the second set, the mapping is not considered a function. These examples are shown at the bottom row of Figure 10.1. In practice, using such mappings means that information is lost. An example commonly occurs in coding when incoming data are not discernable or not able to be matched to a code. The unmapped values are blank in the second set. Thus, although the bottom row mappings are not desirable, they occur in practice due to problems with the data.

The phrase *exactly one* in the definition of a function means that each member of the first set maps to only one member of the second set. This is called single valued and means that one-to-many (last column in Figure 10.1) is not allowed. An example of a one-to-many mapping would be a mistake in the form design creating overlapping age categories, for example, 10–20 years old and 20–30 years old. In the example, a 20-year old would map to both categories. Some cases, however, are not the result of errors,

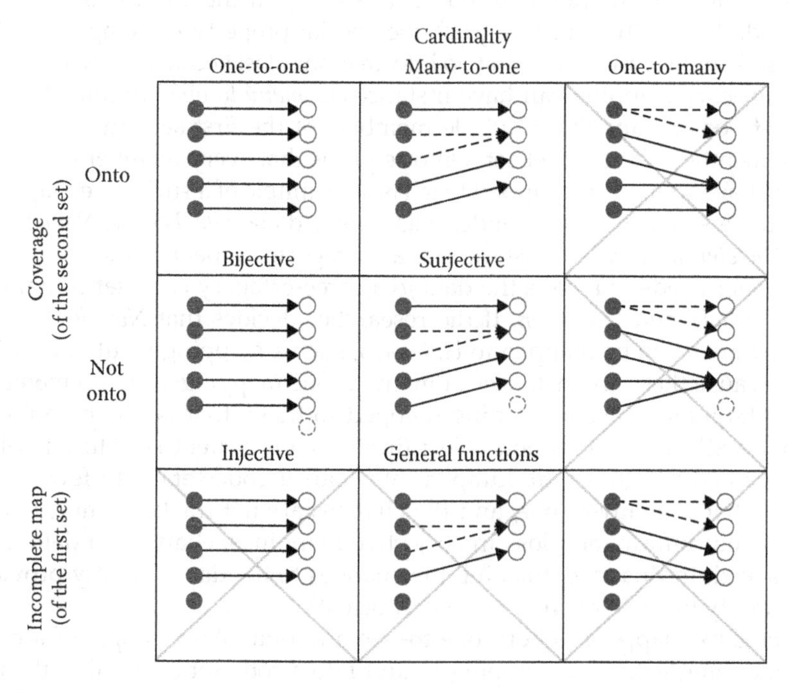

FIGURE 10.1
Mapping possibilities.

for example, a researcher for one study (Study A) collected attribution of adverse reactions as not related, possibly related, and related. The principle investigator of Study B used the following categories: not related, not likely related, likely related, and definitely related. When they tried to combine the data from their studies, the category *possibly related* from Study A could have been mapped to two categories in Study B: *not likely related* and *likely related*. Splitting the data from Study A and calling some of the *possibly related* reactions as *not likely related* and calling others as *likely related* would be questionable if performed after the fact. Using one-to-many mappings is the equivalent of creating information.

An arrangement that meets the definition of a general function is shown in the middle cell in Figure 10.1. Specialized types of functions are in the cells labeled injective, surjective, and bijective. These meet the definition of a function and additional constraints.

An *injective mapping* is a *one-to-one* mapping. Each member of the first set maps to one and only one member of the second set. Consider a data set in which the gender was collected as M for male and F for female, and all animals in the study had a gender recorded. If a researcher mapped the M and F to a three-code set containing the text strings *Male, Female,* and *Unknown,* the mapping would be injective. It is fine that nothing from Study A maps to *Unknown* because injective criterion does not require every member of the target set to be used. An injective mapping has the special property of being reversible. *Reversible* in this case means that no information is lost because of the mapping.

A *surjective* mapping can have instances of *many-to-one* but must be *onto*. Many-to-one means that multiple members of the first set can map to the same member of the target set. Onto requires that every member of the target set have something mapped to it. As an example of a surjective mapping, consider a study in which gender was collected as *Male, Female, Not checked,* and *Undetermined* where instances of all categories appear in the data, and a researcher needs to recode the data to a three-category code set containing *Male, Female,* and *Unknown*. If the researcher decides that *Not checked* and *Undetermined* can be mapped to *Unknown,* such a mapping would be surjective because there are instances of many-to-one mappings, and all members of the target set have something mapped to them, that is, the mapping is onto. Recall that information is lost (information content is reduced) when fine-grained categories are lumped into coarser code sets with fewer categories, that is, many-to-one, and the changes are not tracked. Thus, to prevent actual information loss in surjective mappings, changes to data must be tracked. In this case, tracking the changes to the data not only provides traceability but also prevents actual information loss.

A *bijective* mapping is both one-to-one and onto. An example of a bijective mapping would be mapping M and F to a code set containing the text strings *Male* and *Female* because bijective mappings must also be injective and reversible. Understanding the different types of mapping helps to prevent unwanted information loss in research data.

Formatting Transformations

Formatting transformations are mappings without information loss. Consider an example where gender was collected as a free text field, and responses included *M, F,* and *male, female, woman, Male, Female, N/A, boy,* and blank. Upon seeing the forms, the researcher realized that he or she has made a mistake in not collecting the data in a more standardized way and wants to standardize the data to *M, F,* and *Unk.* There are instances of many-to-one mapping because of the lexical variations in the data. However, semantically, the mappings are effectively one-to-one because the lexical variations *M, male, Male,* and *boy* mean the same thing with respect to the study as do the lexical variations of the concept female. Thus, formatting transformations change formatting only. They do not involve traversing levels of abstraction, variations in meaning, or lumping, or splitting of concepts. Such transformations are usually made to facilitate computation or display of the data. As such, they should be done systematically whether performed algorithmically or manually. Although they do not change the meaning of the data, they do change the representation and thus are often tracked like other data changes. In practice, it is easy for formatting activities to drift into decisions about reducing information content. The two are distinguished here, so that researchers will note the *crossing of the line* when it occurs and will take the opportunity to make deliberate and documented decisions about information reduction.

Conversions and Calculations

A *conversion of units,* for example, converting a length from centimeters to inches, is a calculation that does not change the *dimensionality* of the value. Other calculations, for example, calculating body mass index from height and weight, change the dimensionality of the value. Regardless of this difference, from a data processing perspective, the two can be treated the same. Conversions and calculations usually change data values; thus, like other data changes, they should be systematic, tracked, and free from bias.

Workflow for Transformations

Imputation, standardization, coding, formatting, and calculations are all types of *data transformation.* Transformations can be done manually for small data sets. When this is the case, they are subject to human error and need to

TABLE 10.3

Tasks Associated with Data Transformation

Data Transformation Tasks	All Cases	Special Cases
Specifying the data elements to be transformed	X	
Specifying and documenting the transformation procedure	X	
Specifying the transformation algorithm(s)	X	
Training staff on manual transformations		X
Ongoing review of manual transformations and documentation thereof		X
Programming or otherwise configuring the transformations		X
Testing the transformation algorithm		X
Ongoing monitoring of the transformation algorithm(s) and documentation thereof	X	

be specified ahead of time and assessed on an ongoing basis to assure accuracy and consistency. From a workflow perspective, manual transformations require specification and creation of guidelines, training staff on the guidelines, and ongoing review and feedback to assure consistency (Table 10.3). When transformations are performed by computer programs, they must be specified, programmed, or otherwise configured, and tested (Table 10.3). When testing does not cover each logic condition and variation in incoming data, as is most often the case, the transformations need to be monitored to assure that they are working correctly. From a workflow perspective, automated transformations require specification, programming, and testing as well as ongoing monitoring. Both manual and automated transformations should be tracked at the data value level, that is, when a data value is changed, the value before the change should be preserved; the date and time of the change should be tracked as should the individual or algorithm making the change. Tracking provides traceability from a data value in an analysis data set to the origin of the data.

Data Enhancement

Data enhancement is the association of data values with other usually external data values or knowledge. Examples of data enhancement include geocoding addresses, linking local weather and other environmental data to a study database, coding diagnoses with a medical coding *taxonomy*, and associating lesions on zebra fish with the appropriate structure in the zebra fish *ontology*. There are two varieties of data enhancement. The first is linkage with other data, such as linkage of weather data by zip code described in the farm study example in Chapter 9. Linkage with other data involves the association with more data values via *referential integrity*. As in the farm data example, the associated data values need not be at the granularity of the experimental unit. They may be at a lower or higher level of granularity. Data enhancement of this type is data integration but is usually with external data stores such as environmental data or population data. In human subjects research, linking data may be prohibited by agreements limiting the use of the data or by lack of consent to link data.

The second type of data enhancement is linkage of data elements or data values with external *knowledge sources* such as coordinate systems, taxonomies, or ontologies. In the zebra fish example above, the anatomical zebra fish ontology is a list of anatomical zebra fish structures associated with hierarchical relationships; for example, the tail fin fold is part of the median fin fold, the tail fin fold has parts of ventral actinotrichium, and the tail fin fold is a type of multitissue structure. If a lesion was associated with the tail fin fold and a researcher noticed that the lesion started at the interface between two types of tissue, the researcher could search all such lesions that occurred on any multitissue structure. In this way, the external knowledge allows the researcher to reason over data in new ways. As the number of these external resources increases, the amount of information extractable from data increases. Knowledge sources come in many varieties. Common varieties include *vocabularies*, also called *dictionaries*, terms associated with definitions including very large sets as well as small sets of valid values for enumerated data elements; *taxonomies*, single hierarchy systems; and *ontologies*, multihierarchy systems that store multiple types of relationships between terms as well as logic statements concerning the terms. Data values may be associated with knowledge sources through a coding process (see the coding section above). Association of knowledge sources at the data element level–mapping data elements to some external knowledge source is usually handled as metadata.

From a workflow perspective, data enhancement entails identification of external data or knowledge sources, gaining permission to use them, identifying data elements to be associated or that can serve as link fields, reconciling the data to check for any linkage problems, and ongoing monitoring of linked data to check for problems. Data elements that are good link fields are those that are complete (have a very low number of missing values), are unique themselves or in combination with other data elements, and are accurate. Link fields are sometimes called matching fields. Reconciliation is the process of assessing linked data to identify problems in the linkage. For example, if data from a clinical lab were matched with a study database by the participant number and time point (the link fields), reconciliation might entail looking to see that the gender and age received from the lab match the gender and age in the study database (Table 10.4).

TABLE 10.4

Tasks Associated with Data Enhancement

Data Enhancement Tasks	All Cases	Special Cases
Identification of external data or knowledge sources	X	
Gaining permission to use external data or knowledge sources	X	
Identifying data elements that can serve as link fields		X[a]
Specifying process for associating external knowledge to study data		X[b]
Specifying, programming or configuring, and testing record linkage algorithms		X[a]
Ongoing reconciliation of linked data to check for any linkage problems		X[a]
Specifying, programming or configuring, and testing algorithms to associate data with external knowledge		X[b]
Manual association of data that could not be automatically associated		X[b]
Ongoing review of manual association and documentation thereof		X[b]

[a] External data only.
[b] External knowledge sources only.

Summary

This chapter covered common data processing operations performed on research data. Three trends run through all of the sections in this chapter. The first is that operations that change data should be tracked such that any changed value can be traced back to the original value, and the algorithms or other processes that changed the data. As mentioned in Chapters 6 through 8, computer programs that make such changes need to be supported by documentation that they work correctly. A second trend that permeates the chapter is the necessity for consistency. All changes to data should be systematic. Where manual, that is, human-performed, tasks are included in data processing, steps should be taken such as process standardization and training to assure consistency; consistency should be measured, and the measurements used to monitor the process provide feedback and control consistency. The third trend running through the chapter is that efforts to standardize and increase consistency can introduce information loss or bias. For this reason, a statistician familiar with the planned analysis should be included in all decisions that alter data. These three concepts, traceability, systematicness, and freedom, from information loss and bias substantially increase the likelihood that research is reproducible.

Exercises

1. A researcher is receiving data for a study from an international commerce bureau. The data are for the prior five calendar years and include commerce metrics provided to the bureau by each country. Which of the following data cleaning strategies is the most appropriate?
   a. Identify and flag
   b. Identify and notify for future
   c. Identify and resolve
   d. None of the above

   Correct answer: (a) Identify and flag. The discrepancies will not likely be fixed if the researcher seeks to resolve or to notify because they are national statistics over which the investigator has no control.
2. In the scenario above, the researcher plans to link the commerce metrics to weather patterns by country and test a hypothesis about weather patterns impacting global commerce. What would this type of data integration be called?
   a. Data standardization
   b. Data coding
   c. Data enhancement
   d. Data imputation

   Correct answer: (c) Data enhancement. The researcher is linking study data to an external source of data.
3. In the scenario above, which choice below is the most appropriate linkage field?
   a. Zip code
   b. Country code
   c. Country name
   d. Postal region

   Correct answer: (b) Country code. Country name may be free text and not standardized; if so, it would be susceptible to lexical variation that would hinder linkage.
4. In the research scenario above, if the country names were entered as free text, which process below should be applied to the data?
   a. Data standardization
   b. Data coding
   c. Data enhancement

d. (a) or (b)
e. (a) or (c)

Correct answer: (d) (a) or (b). The free text data could be coded with country codes, or the data could be standardized. In this example, a and c are effectively the same thing because data would likely be standardized to the codes.

5. Data from an agroforestry study are collected at five farms on paper data collection forms. After each quarterly assessment, the forms are sent to the researcher where they are entered. Which data cleaning strategy is the most appropriate?
   a. Single data entry with on-screen checks
   b. Single data entry followed by discrepancy identification and resolution
   c. Double data entry with on-screen checks
   d. Double data entry followed by discrepancy identification and resolution
   e. None of the above

   Correct answer: (d) Double data entry followed by discrepancy identification and resolution. Double data entry decreases random keystroke errors, so they will not be mistakenly identified and sent to the farms for resolution. On-screen checks are not appropriate because data are not entered at the farms.
6. In the farm study above, genus and species of parasites found on the trees are identified and written on the form by a data collector. The data are coded by an entomologist at the central center. She looks up the correct spelling and writes the genus and species out in a standardized way prior to data entry. A 10% sample of the coding is recoded by a second person for quality control. What potential source of errors has not been addressed?

   Correct answer: The parasite identification by the data collector is also a manual process and requires quality assessment and control.
7. The researcher in the above agroforestry question wanted to screen the species data for potential inaccuracy. He or she has a list of the parasite species known to exist in the region. He or she writes a rule in his or her data system to flag any species not in that list. Which of the following did he or she leverage?
   a. Comparison to a source of truth
   b. Comparison to a known standard or expectation
   c. Comparison to an independent source of the same information
   d. Comparison to an upstream version of the same information
   e. Comparison to other values in the same study

Correct answer: (b) Comparison to a known standard or expectation. The list of species known to exist in the region is the general expectation for the species that should be found.

8. In another but similar agroforestry study, the researcher had the data collectors collect samples of each distinct parasite identified on each of the trees in the study plots. The samples were labeled with the tree and plot number and sent to a central lab, where the parasites were classified by two independent people. Which of the following comparisons was used?
   a. Comparison to a source of truth
   b. Comparison to a known standard or expectation
   c. Comparison to an independent source of the same information
   d. Comparison to an upstream version of the same information
   e. Comparison to other values in the same study

   Correct answer: (c) Comparison to an independent source of the same information. The two classifiers at the central lab represent two independent sources of the same information.

9. Of the two processes described in questions 6 and 8, which is more rigorous and why?

   Correct answer: Collecting the actual parasites from each tree and sending them to a central center for double independent classification. The event of interest is preserved, and the classification is confirmed through having it done twice.

10. A researcher collects data on a one-page form with 26 data elements. He or she later enters the data into a spreadsheet. Describe a way that traceability could be maintained through his or her planned data standardization, coding, and cleaning processes.

    Correct answer: There are multiple ways this could be done. One way is to record all changes to data on the data collection forms and make the changes in the spreadsheet with no additional tracking in the spreadsheet. In this case, the spreadsheet will reflect the final values, and the paper form is the record of all changes to data. A second way would be to add a tab in the spreadsheet where copies of records prior to changes are saved along with the date of the change and person making the change.

References

ASQ Quality Costs Committee. *Principles of Quality Costs: Principles, Implementation, and Use*. 3rd ed. Jack Campanella (ed.). Milwaukee, WI: ASQ Quality Press, 1999, pp. 3–5.

Boehm BW and Papaccio PN. Understanding and controlling software costs. *IEEE Transactions on Software Engineering* 1988; 14(10):1462–1477.

Kush RD, Bleicher P, Kubick W, Kush S, Marks R, Raymond S, and Tardiff B. *eClinical Trials: Planning and Implementation*. Boston, MA: Thompson Centerwatch, 2003.

Ross JE. What is the 1-10-100 Rule? Total Quality Management, February 2009. Accessed August 28, 2012. Available from http://totalqualitymanagement.wordpress.com/2009/02/25/what-is-1-10-100-rule/.

Spiceland JD, Sepe JF, and Nelson MW. Intermediate Accounting eBook th ed., Chapter 20: Accounting Changes and Error Corrections. McGrawHill, 2010. Available from http://connect.mcgraw-hill.com.

Walker B. The real cost of bad data, the 1-10-100 Rule. A Melissa Data White Paper. Accessed August 28, 2012. Available from www.melissadata.com/dqt/1-10-100-rule.pdf.

11

Designing and Documenting Data Flow and Workflow

Introduction

As decisions about data entry and processing are being made, it is helpful to show the data and workflow graphically. Graphical diagrams help communicate data sources and flow and the tasks involved in data processing. Creating the diagrams helps identify what processing may be needed, and after the study, the diagrams, at a high level, document what was done. This chapter describes two common diagramming notations and prepares the reader to create data flow and workflow diagrams for research projects.

Topics

- Importance of diagrams
- Diagrams as models
- Data flow diagrams
- Workflow diagrams

Importance of Diagrams

Before written language, early humans used symbols to communicate. With graphic representations, humans perceive more information and perceive the information faster than through verbal and written communication channels (Wickens and Hollands 1999). Humans directly perceive meaning through symbols. For example, the skull and cross bones is a universal symbol for danger. Red, yellow and green lights signal drivers when to stop or proceed, and a road sign with a picture of a tipping truck lets drivers know that the road conditions are conducive to overturning a vehicle. Symbols are used to quickly communicate meaning.

For these reasons, standard symbols are used to clearly communicate process information, for example, the steps and their order in a workflow or data flow. The symbols communicate meaning, much like road signs. Process diagrams often are drawn at a level of detail that enables them to fit on one or a few pages. As such, they provide a way to, in one place, see the whole process. Different types of diagrams show different aspects of processes. The workflow diagrams (WFDs) and data flow diagrams (DFDs) recommended here support data management planning and documentation by showing data sources, process steps and their sequence, the roles involved, context, data flow, and transformations (Figure 11.1). There are other process aspects that are not critical to data management planning in research and thus will not be discussed in detail here.

The act of documenting a process requires each step to be understood and made explicit. Making the process steps and their sequence explicit aids in data management planning, especially by making the process easier for others to understand and critique. A graphic depiction of all data sources and how they are processed facilitates identification of overlooked data or invalid assumptions about the need (or not) for specialized data processing. In the author's experience, several instances of aided planning through flowcharts come to mind, as do several examples in which diagrams were not used where external data sources were missed and not known to the research team until requested for an analysis. While on the surface diagramming a process or data flow seems simple; many processes are complex and can be examined from different perspectives, including static and dynamic aspects, data flow versus workflow, and at different levels of detail. The time invested in WFDs and DFDs is well spent because they are among the most used and shared artifacts produced in the operationalization of a research design.

Process aspects depicted	Data flow	Workflow
Context	X	
Process steps/tasks	X	X
Data content		
Data transformation	X	X
Sequencing/control		X
State		
Roles		X

FIGURE 11.1
Process aspects covered by DFDs and workflow diagram. (Workflow diagrams described here use conventions described in the ISO 5807 standard. DFDs described here use Yourdon symbols and conventions.)

Diagrams as Models

Pictures are usually provided when something is being sold on line. However, to show how something works, a diagram of the inner mechanical parts and their relationship to each other is probably the best depiction. Similarly, to look for potential explosive and other harmful objects, airport security uses X-rays to view the contents of luggage and packages rather than video images of the outside of the bags. Importantly, each perspective, inner working, and outward appearance are a model of the actual item; each represents different aspects of the item. A photograph or painting is one model of a house. A set of drawings is a model of different aspects of the house and is also used for a different purpose such as a blueprint for building the house. The rendering and the blueprints are at the same time abstractions and templates depending on how they will be used. Workflow diagrams and DFDs are abstractions because they show only certain aspects of workflow and data flow. They are also templates where they are used to govern processes and to guide how data processing systems are set up.

A process can be composed of tasks accomplished by both human and machine. Humans can perform both physical and mental steps or tasks, whereas machines can only perform physical and computational logic tasks.

Computers are commonly used for data processing tasks. Although the computer performs the physical (logical) manipulations of the data, interpretation and thinking remain the role of the human. Thus, humans interact with computers both in physical workflow and in data flow. Further, we make machines perform human like tasks by reducing human thought into a set of conditions that a machine can recognize and a set of logical instructions that a machine can carry out under the conditions.

Process diagramming is complicated by the intertwined nature of humans and computers in data processing. Further, some researchers may use humans for data processing tasks, whereas others may use computers for the same tasks. One approach to dealing with this complexity is to use different representations of processes, one that concentrates on process steps and another that concentrates on data flow.

Two Types of Diagrams

The distinction between workflow and data flow is sometimes blurry. However, the distinction is important because the underlying thing being represented, that is, process steps or tasks versus movement of data, is different. Data flow is about the data points that are being communicated or transferred, where the data are stored, and how those data are transformed.

However, for workflow, we care about the physical and sometimes mental steps that occur and the order in which they occur. Workflow is the steps or tasks performed by humans or computers and their sequence. In research settings, processes have both workflow and data flow components that need to be represented. Sometimes, the emphasis on one or the other is less, and one representation can be used; often, both are important and multiple diagrams are required. Use of one or the other or both diagram types is common. However, blurring the two different concepts of data flow and workflow in one diagram usually provides an incomplete representation of both the data flow and the workflow and is not recommended. It is important for the researcher to be clear and to deliberately make decisions about both data flow and workflow.

Flowcharts are the most common type of workflow diagram and are easily understood by most people; thus, they are widely used. Most of the symbols needed to create flowcharts are in common use in quality improvement work and are included in word processing, drawing, and presentation software packages, such as the Microsoft Office packages. This chapter applies to research this common flowchart notation that ironically has its history in the ISO 5807 information processing standard. DFDs are less widely known and have been used mostly in the information systems and data processing communities. Yourdon notation is a set of symbols and conventions for DFDs named for the person who developed it, Edward Yourdon. This chapter applies Yourdon notation to the graphical depiction of data flows for scientific studies.

Data Flow and Context

Data flow is the movement of data. *Data flow diagrams* (also referred to as DFDs) provide a way to document and visualize the movement of data and operations on data. They show sources of data and places where data are stored. The latter are called sinks. Like workflow diagrams, DFDs can be drawn at different levels of detail: a high-level DFD that is called a context diagram (Figure 11.2) and the collection of more detailed versions called DFDs (Figure 11.3). Both use the same notation; the only difference is that the context diagram is a high-level view of how a study, a system, or an organization interfaces with the outside world rather than the details of how data move and are processed within a study, a system, or an organization. A context diagram is drawn with respect to a system, a study, or an organization and shows all data sources and sinks and the major operations performed on data between the sources and the sinks. A context diagram shows the context or environment in which the system, study, or organization operates through data flow relationships with external entities, processes and data stores. Thus, context diagrams are helpful to document and share knowledge about the data-related scope of a study, that is, what data sources, sinks, and operations are included in a study. For example, context diagrams may show important

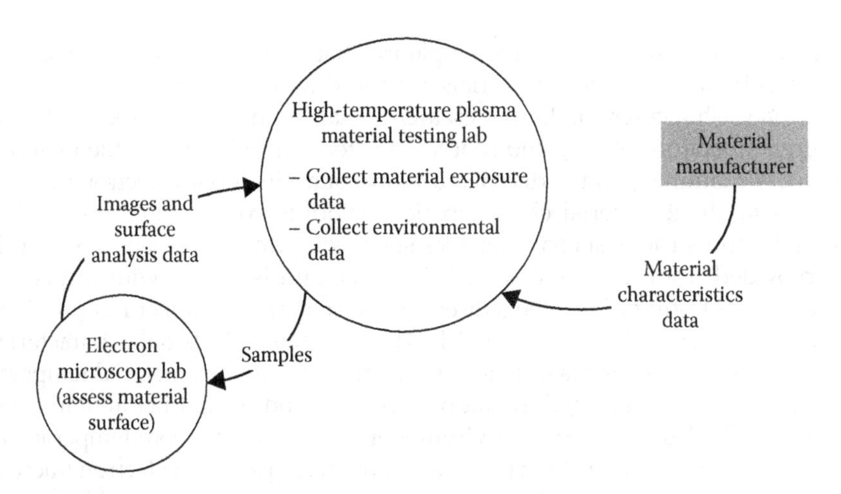

FIGURE 11.2
Example context diagram.

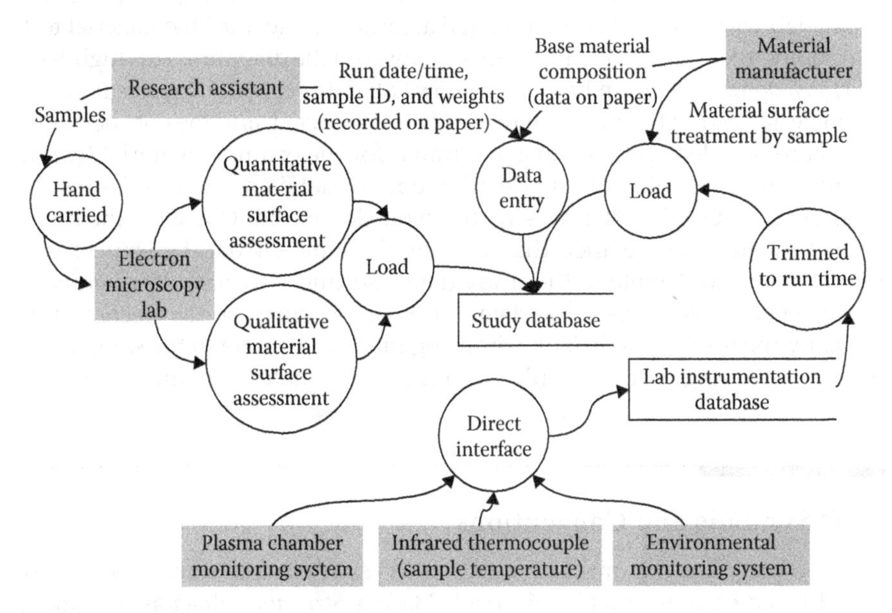

FIGURE 11.3
Detailed DFD.

data exchanges between academic collaborators or may depict the major data sources and operations performed on data as an overview for the research team, collaborators, or research sponsors. Context diagrams are the first diagrams that should be developed because they explicitly show all data sources and interfaces between the study and the outside world, and show everything under consideration from a data perspective at a high level (Yourdon 1989).

The plasma trash incinerator example in Chapter 5 uses a context diagram (Figure 5.4). In the example, materials are tested for a high-temperature use in a plasma trash incinerator. In the researcher's lab, each material is subjected to a high-temperature plasma and sent to an external facility where the material surface is examined, and surface features are quantified with electron microscopy. Data about material characteristics including composition, production methods, and structural characteristics such as hardness and tensile strength are provided from the manufacturer. The diagram is drawn with respect to the materials testing lab. An outside entity, the material manufacturer provides material characteristics data to the lab. The lab itself collects data characterizing the exposure of the material to the plasma, for example, material temperature, plasma temperature, duration of exposure, and weight before and after exposure. The lab also records environmental data such as room temperature, humidity, date, and time of the material exposure. An external electron microscopy lab evaluates the samples before and after exposure and provides quantitative information such as microcrack count, crack length, and depth, as well as a qualitative description of the material surface. The data are provided to the testing lab where the data are stored (all arrows point toward the material testing lab, and there are no other data sinks shown on the diagram). This high-level DFD is a context diagram because it shows all data-related interfaces between the material testing lab, in which the study is conducted, and the outside world.

A more detailed DFD can also be drawn for the example material testing study (Figure 11.3). In gaining detail, the more detailed diagram looses the clear depiction of the interfaces with the outside world. However, the detail gained in the more detailed diagram clearly depicts what the sources and types of data contribute to the study database, and indicates the operations that are needed to process the data, for example, entry of data recorded on paper by the research assistant, trimming monitoring data to the sample run time, and loading of the data file provided by the material manufacturer.

DFD Symbols and Conventions

The two diagrams (Figures 11.2 and 11.3) use symbols and conventions introduced in Edward Yourdon's 1989 book *Modern Structured Analysis*. Variations of this notation exist, but the symbols and conventions used here are sufficient for most research applications. The symbols used in Figures 11.2 and 11.3 and enumerated in Figure 11.4 include closed boxes, open boxes, circles, and connectors with arrows. Briefly, the closed boxes (shaded here to more clearly distinguish them) represent *entities* that are origins or consumers of data. Entities are external to the processes (circles) in the diagram. For example, in Figure 11.3, the research assistant provides samples to the electron microscopy lab but is not part of that lab's process. The samples are an input to the qualitative and quantitative analysis processes performed by the electron microscopy lab. Terminators

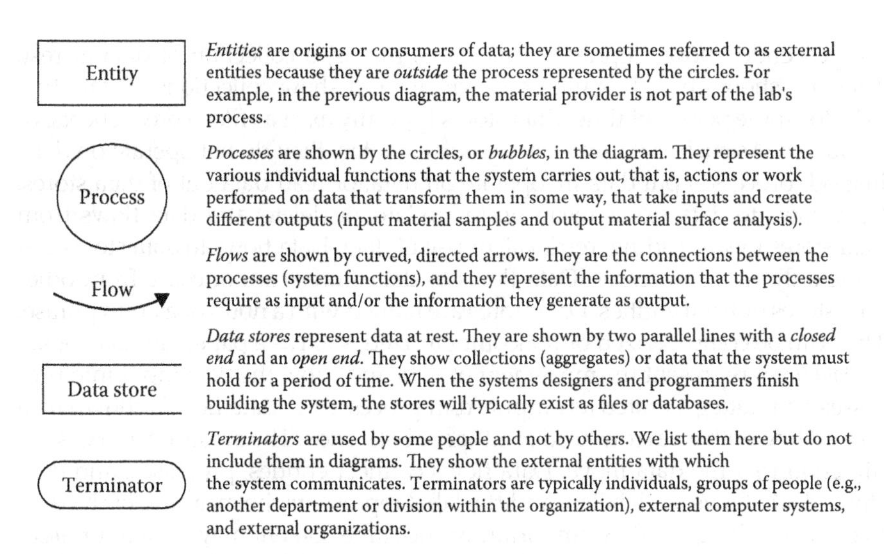

Entity
Entities are origins or consumers of data; they are sometimes referred to as external entities because they are *outside* the process represented by the circles. For example, in the previous diagram, the material provider is not part of the lab's process.
Process
Processes are shown by the circles, or *bubbles*, in the diagram. They represent the various individual functions that the system carries out, that is, actions or work performed on data that transform them in some way, that take inputs and create different outputs (input material samples and output material surface analysis).
Flow
Flows are shown by curved, directed arrows. They are the connections between the processes (system functions), and they represent the information that the processes require as input and/or the information they generate as output.
Data store
Data stores represent data at rest. They are shown by two parallel lines with a *closed end* and an *open end*. They show collections (aggregates) or data that the system must hold for a period of time. When the systems designers and programmers finish building the system, the stores will typically exist as files or databases.
Terminator
Terminators are used by some people and not by others. We list them here but do not include them in diagrams. They show the external entities with which the system communicates. Terminators are typically individuals, groups of people (e.g., another department or division within the organization), external computer systems, and external organizations.

FIGURE 11.4
DFD symbols.

are a special type of entity designating external entities with which the system communicates. In Figures 11.2 and 11.3, the material manufacturer would be considered an external entity. In the context diagram (Figure 11.2) drawn from the perspective of the material testing lab, the electron microscopy lab is considered an external entity. However, if the perspective is that of the research study, the electron microscopy lab would likely be considered as an internal entity because it is clearly part of the study. An entity is named with a noun or a noun phrase. Data flows can come to and from entities and only from processes.

Processes are shown by the circles, or *bubbles*, in the diagram. They are the various steps, tasks, or operations that are carried out; in particular, processes are actions or work performed on data that transform them in some way. They take inputs and create different outputs, for example, the data recorded on paper is converted to electronic data in the study database by the data entry process, and the external data files provided are integrated with data in the study database during the load processes. As such, a process meets the definition of a system and may be referred to as a system. A process is named or described with a single word, phrase, or simple sentence that describes *what* the process does. A good name consists of a verb–object phrase such as assessment or assess material surface. In some cases, the process will contain the name of a person, a group of people, a computer, or a mechanical device that performs the process. For example, in Figure 11.2 context diagram, the electron microscopy lab was depicted as a process and named *electron microscopy lab* to represent the analysis performed by the lab. That is, a process is sometimes described by who or what is carrying out the process, rather than describing what the process is. A process must have both inputs and outputs. The processes may be numbered by placing a unique number at the top or bottom of the circles.

Open ended boxes represent *data stores*, that is, a collection of data at rest. They are shown by a box with an *open end* and show collections of data that exist for some period of time. Data stores typically exist as files, spreadsheets, or databases. As such, data stores are passive, that is, they do not operate on data, instead, processes put data in, operate on data, or read data out of data stores. Data flows to data stores mean write, update, or delete; and data flows from data stores mean reading, retrieval, or use of data. Data flows to data stores can come only from processes. Data flows to data stores cannot come from other data stores or from entities. Data stores are named with a noun or a noun phrase. They can be computerized or noncomputerized, such as a paper lab notebook.

Data flows represent the movement of data and show the data input into processes and data generated as output from processes. A data flow is shown by a curved line with an arrow head at either or both ends. The arrow designates the direction that the data move. Data flows connect entities, processes, and data stores. Data flows are labeled, and the label represents the meaning of the data that moves along the flow. Importantly, the same content may have a different meaning in different parts of a diagram, for example, an address as given by a research subject versus an address that has been matched and validated.

An event list often accompanies detailed DFDs. It contains things that stimulate action from the system. For example, in Figure 11.3, the research lab placing an order for samples may trigger the receipt of the material composition data in the shipment with the samples and may also prompt the manufacturer to send the electronic file with the material surface treatment information.

A good DFD fits on one page, is not too crowded, is as simple as possible, and depicts the necessary entities, processes, stores, and flows. Where additional details are needed, processes can be drawn in greater detail on a new page; however, in many research settings, a simple context diagram or a context diagram and one more detailed diagram such as those shown in Figures 11.2 and 11.3 will suffice. It is generally best to start with a higher level context diagram to understand the scope and boundaries and to subsequently decompose processes to lower levels of detail only when needed. Additional diagrams may be needed in research settings where many different types of data exist, where data movement is extensive, or where information systems must be built or customized for the research. The diagrams are clearer if everything on one page is depicted at the same detail level. The optimal detail level for a diagram is not always evident. Often, it is helpful to add entities, processes, and data stores as they come to mind or as they are decided in a research design and save decisions about lumping lower detail levels into a higher level process, entity, or data store, or splitting higher level processes, entities or data stores into separate and more detailed ones for later iterations.

The size and shape of bubbles are generally up to the diagram creator. Curved or straight arrows can be used, but only one or the other should be used because the diagrams are visually clearer with one or the other. As a simple rule from cognitive science, any distinctions made by the diagrams should be meaningful. For example, if dashed lines and solid lines are both used, the dashed lines should mean one thing, whereas the solid lines mean another. Similarly, some people use color to differentiate types of entities or flows. It may take the researcher several tries before a clear and complete diagram is achieved. To this end, Table 11.1 provides a list of conventions for creating Yourdon-style DFDs. Diagrams can be improved by reviewing each iteration against the conventions.

TABLE 11.1

Conventions and Rules for DFDs

Entities

Closed boxes represent entities.
Entities are producing, providing, or consuming data.
An entity is named with a noun or a noun phrase.
Data flows can come to and from entities ONLY from processes.

Processes

Processes are represented by circles or *bubbles*.
Processes are steps, tasks, or operations that the system carries out.
Processes have both inputs AND outputs.
A process is named with a single word, phrase describing *what* the process does.

- A verb–object phrase such as assessment or assess material surface.
- A role, group of people, computer, or mechanical device that performs the process.

Processes may be numbered.
Processes may be accompanied by an external description of how the process is initiated.

Data Stores

Data stores are represented by a box with an *open end*.
A data store is a collection of data at rest, for example, files, spreadsheets, lab notebooks, or databases.
Data flows to data stores can ONLY come only from processes.
Data flows to data stores mean write, update, or delete.
Data flows from data stores mean reading, retrieval, or use of data.
Data flows to data stores cannot come from other data stores or from entities.
Data stores are named with noun or noun phrase.

Data flows

Data flows are represented by a line with an arrow head at either or both ends.
A data flow is the movement of data.
An arrow head designates the direction that the data move.
Data flows connect entities and data stores to processes.
Data flows are labeled.

Additional Checks for Diagrams

Processes that have outputs but no inputs (miracles[a])
Processes that have inputs but no outputs (black holes[a])
Unlabeled flows, processes, entities, and stores (mysteries[a])
Undirected flows, that is, flows without an arrow head (flippers)
Disconnected entities and data stores (floaters)

[a] The labels' miracles, black holes, and mysteries were coined by Edward Yourdon and are retained here because they help practitioners recall error-checking strategies.

Workflow Diagrams

Workflow is the sequence of steps in a work process. Workflow diagrams are often called process diagrams or flowcharts. A workflow diagram (process diagram or flowchart) is a graphic depiction of the steps or activities that constitute a process. Indeed, each process bubble in a DFD could have a workflow diagram supporting it. Various definitions abound for these terms; some fail to distinguish between data flow and workflow. Work tasks and movement of data may coincide within a series of process steps; data flow and workflow are treated separately here using separate DFDs and workflow diagrams because in a research study, each must be clearly articulated and combining the two risks incomplete treatment of either or both. Thus, data may appear on a workflow diagram, but the focus of the diagram is the tasks performed and their sequence rather than the movement of data. Figure 11.5 shows an example workflow diagram for the data collection process in the material testing study for which context and data flow are shown in Figures 11.2 and 11.3, respectively.

The material testing workflow diagram (Figure 11.5) shows the steps involved in conducting the material tests. The process is initiated when a sample is selected for a scheduled test. The selected sample is cleaned, weighed, and then sent for the pretest electron microscopy. After the electron microscopy is complete, the sample is placed in the plasma chamber; the chamber is closed, and baseline data are acquired to assure that the instrumentation correctly reads ambient values. If the instrumentation readings differ from ambient values, the test is aborted. If the instrumentation readings match the ambient readings from the lab environmental monitoring, the test is conducted. Simultaneously, the test environment (the plasma) is generated to match test design, and during the test, data are acquired from the plasma chamber monitoring equipment. The posttest chamber data acquisition continues until the material temperature equalizes with ambient readings. Once the tested material has equalized, the chamber is opened and the sample is removed, weighed, cleaned, then reweighed, and sent for posttest electron microscopy. The sample test is considered complete following the posttest electron microscopy.

Like DFD, workflow diagrams are constructed from standard symbols. The symbols communicate specific meaning; for example, a line with an arrowhead communicates the direction of data flow. Thus, knowing the meaning of the symbols is important in understanding and creating the diagrams.

Workflow diagrams are sometimes used to show process steps categorized by the roles that perform or have responsibility for them. From a project management perspective, this can be helpful in clearly showing who or what organization is responsible for which tasks. Roles are shown on the diagrams in rows or columns depending on the horizontal or vertical

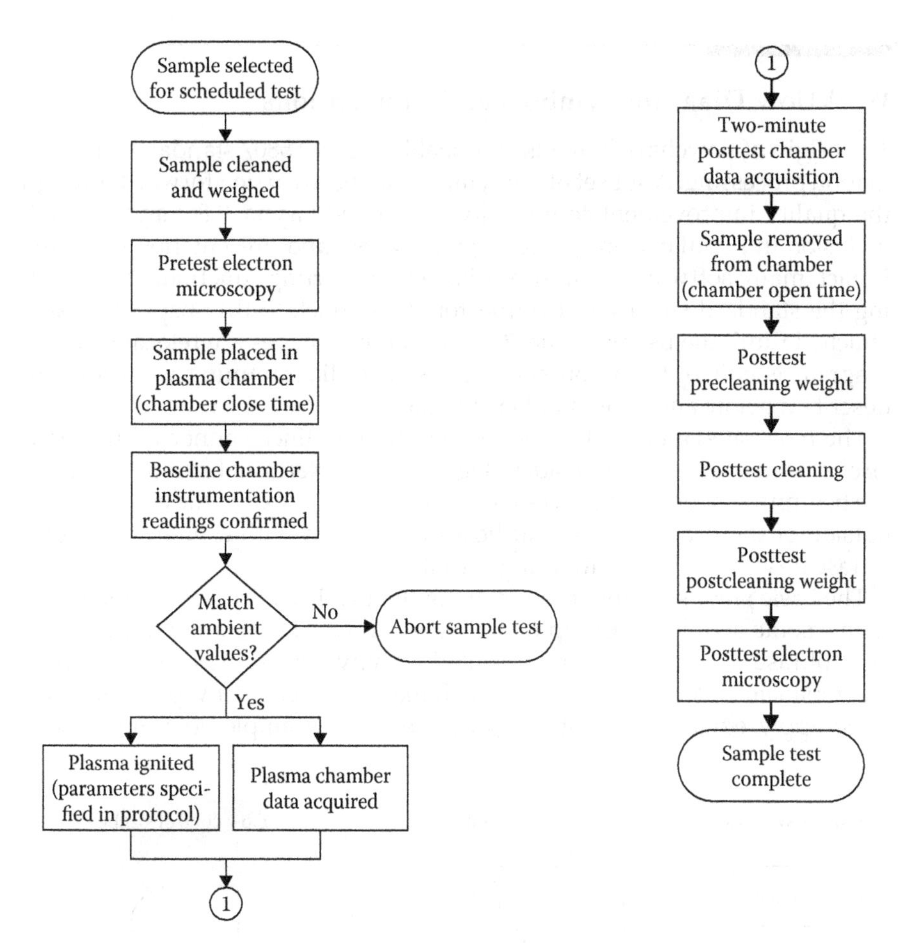

FIGURE 11.5
Example DFD.

orientation of the diagram. The rows or columns are called swim lanes. Process boxes are placed in the lane according to the role that performs them. Workflow diagrams with swim lanes can be helpful on studies with multiple collaborators or collaborating organizations performing different process steps. Adding role information to a workflow diagram can make the diagram more complex and the visual depiction of steps and their sequence less obvious. Whether the advantages outweigh the disadvantages depends on the complexity of the diagrammed process and the necessity of clarity around roles. As an alternative, where task-based role information is important, a workflow diagram without swim lanes can be accompanied by a task responsibility matrix that lists process tasks down the left-hand column and roles or organizations across the top and uses the cells to assign tasks to roles or organizations.

Workflow Diagram Symbols and Conventions

The origin of flowchart symbols is probably the ISO 5807 standard for information processing. A subset of the symbols has been carried forward through the quality improvement community (Juran and Gryna 1988) and through inclusion in popular word processing, drawing, and presentation software. In fact, many software applications have flowcharting functionality including the standard shapes and connectors that attach to the shapes and stay attached (until the user detaches them), whereas shapes are relocated as the diagram is edited. Common symbols used in diagramming workflow are described below and shown in Figure 11.6.

The *terminal* symbol is two horizontal parallel lines connected by semicircles on the left and right sides (Figure 11.6). A terminal symbol signifies the beginning or end of a process or an entry from or exit to the environment outside of the process. Terminal boxes are named with a noun or a noun phrase describing the terminating activity.

The basic *process* symbol is a rectangle (Figure 11.6). Process symbols designate some activity or group of activities. Process boxes are named with a verb phrase describing the represented activity, for example, identify data discrepancies. Decisions made in activities represented by process boxes must apply equally to all items processed. For example, identification of

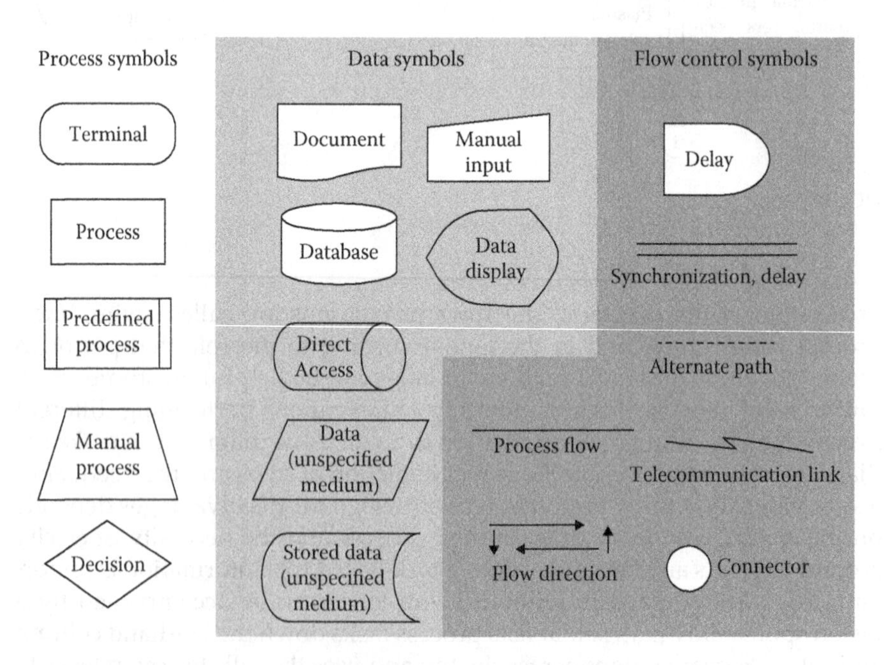

FIGURE 11.6
Workflow diagram symbols.

data discrepancies may associate a decision, discrepant, or not discrepant with each item assessed. Differential action impacting the sequence of steps immediately following the process box is not permitted in a process box; action dictating flow control or sequence of steps can only be represented in a decision box.

There are three special types of processes each with its own symbol (Figure 11.6). A predefined process is a process that is specified or defined elsewhere. A predefined process is represented by a process rectangle with double lines on the vertical sides. An example of a predefined process in research would be a lab protocol or a standard operating procedure. The second type of special process is a manual process. A manual process is one that is performed by a human rather than a machine. Where distinction between human and automated tasks is important, the manual process symbol should be used. A manual process is represented by a trapezoid.

A decision is the third type of special type of process. The *decision* symbol is a diamond (Figure 11.6). Decision symbols represent a decision point from which the process branches into two or more paths. For example, the *identify data discrepancies* process might be immediately followed by a decision box labeled, *Discrepant?* where, depending on whether a data value is discrepant or not, a different path is followed. The path taken depends on the answer to the question appearing within the diamond. Each path is labeled to correspond to an answer to the question. The decision can logically be thought of as a switching function having a single entry and multiple possible exits, one and only one of which may be activated following the evaluation of condition (question) defined within the symbol. Decision symbols can show Boolean decisions (yes/no, true/false) or decisions with multiple possible outcomes.

There are two ways of representing a delay in a workflow. The first uses a *bullet* symbol similar in appearance to and connoting a logical *and gate* to represent a point in a process where two or more tasks that occur at least partially simultaneously in time must all complete before the process can continue. Graphically, the inputs connect on the straight vertical side and the outgoing flow emanates from the curved side. An alternate depiction entails the use of two horizontal parallel lines called synchronization lines. The parallel lines denote that tasks above them must be completed or otherwise come to the denoted state before tasks below them can be started. Consider a drive-through window at a fast food establishment. The customer must pay for the order, and the order must be ready before it can be given to the customer. Figure 11.7 shows the two alternatives for diagramming a process delay.

The *document* symbol is a box with unequal sides closed by a curved line (Figure 11.6). It represents human-readable data on any medium used or produced by a process. A document can be an input to or an output from a workflow. Documents are not tasks or actions; thus, no outflow is possible

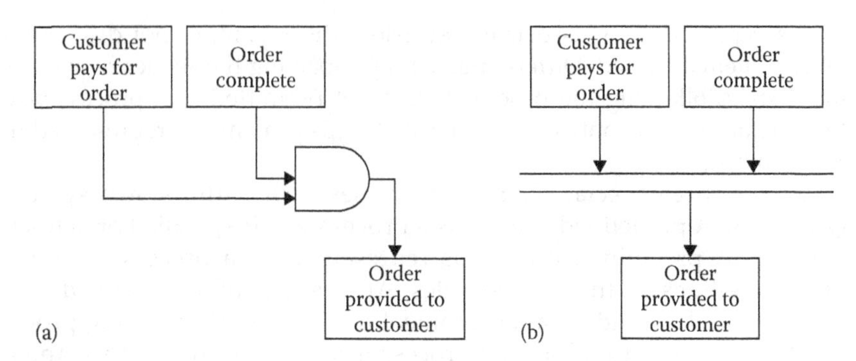

FIGURE 11.7
Representing workflow delays. (a, b) Bullet symbol representing process delay.

from a document. An output document symbol is usually immediately preceded by a task that generates the document. An input document is immediately followed by a task in which the document is used.

There are multiple symbols for different types of data likely stemming from diagramming in early information processing applications. In the early days of computing, how the data were stored was very important. Many storage mechanisms (punch cards, magnetic tape, etc.) were difficult to access; thus, distinguishing between different data formats was informative. In the ISO 5807 standard, there are symbols for many different types of storage. Today, data are much simpler to access and more often are stored in ways that they can be accessed easily. Today, for the purposes of diagramming data flow, it is most important to distinguish between data that are accessible to the process or user, and data that are not, and for what reason. The symbol that is used most frequently today to represent *data storage* is a cylinder (Figure 11.6). The cylinder represents a database. Vertical cylinder is commonly used today to represent storage of data in a database. It is not listed in ISO 5807 but is often confused with the horizontal cylinder. Horizontal cylinder, one of the older symbols in ISO 5807, is used to represent *direct access storage* of data. Direct access storage can be thought of like a book with a good index that enables a user to quickly go to any topic in the book. This is much faster than a book with no index where you have to read through the whole thing to find the topic of interest.

Two basic symbols for data described in the ISO 5807 standard include the parallelogram with two nonperpendicular sides representing any data with the medium being unspecified and a parallelogram with two curved sides representing stored data (medium unspecified) in some format suitable for processing (Figure 11.6). There are also specific symbols representing data stored internally in a computer program or system, sequential access data, and data stored on punch cards and punched tape. Symbols denoting data displayed on a computer screen and manual input of data at the time of processing are also available.

Process *flow lines* represent a path between process elements. A symbol occurring after another occurs sequentially in time after the preceding symbol. A solid straight line is used to denote a process flow (Figure 11.6). A dashed line is used to denote an alternate path. A jagged line is used to denote data transfer by a telecommunications link. Diagonal lines and curved lines are not used on workflow diagrams. Arrowheads represent the direction of the flow on all lines with the exception of synchronization lines. There are no arrows on synchronization lines. All lines that represent flow based on decisions should be labeled. Table 11.2 enumerates conventions for workflow diagrams.

One final symbol used in workflow diagrams is the *connector* symbol. The connector symbol is used to facilitate continuation of process flow when a diagram has an exit to or an entry from, another part of the same diagram, such as continuing the diagram across pages, or to break a line and to continue it elsewhere to reduce visual clutter on a diagram. A circle containing a number, letter, or some other unique identifier represents such a continuation (Figures 11.5 and 11.6). The identifier prevents confusion where more than one such continuation occurs within the same diagram. The workflow diagram in Figure 11.5 shows use of a connector.

TABLE 11.2

Conventions and Rules for Workflow Diagrams

Process symbols
Process boxes must have an input and an output.
Exit terminal boxes must not have output.
Decision diamonds usually represent Boolean decisions and have two corresponding outward paths; both should be labeled.
Decision diamonds with more than two outward paths should all be labeled.
Data symbols
Data boxes including documents as input to processes should be immediately followed by a process box.
Data boxes including documents as output from processes should be immediately preceded by a process box.
Process flow
Flow lines should not cross; use connectors or loop overs to prevent crossing lines.
Diagrams should read from top to bottom or left to right.
All forward progresses should be in the same direction.
All rework or other backward progresses should be in the direction opposite that of forward progress.
All flow lines should be straight or straight with 90° angles.
Diagonal lines are not permitted.
Curved lines are not permitted.
Additional conventions
All boxes should be the same size.
All font in boxes should be the same size.
Diagrams should be uniform with respect to level of detail depicted.

Workflow diagrams should *read* from top to bottom or right to left such that all forward progress of the workflow occurs in the same direction with only rework or other regression occurring in the opposite direction. Items shown on the same diagram should be at the same detail level. Combining items at different detail levels can cause confusion. Text descriptions in symbols should be as short as possible, so that the diagram is more readable. Annotations, also called callouts, can be added if necessary, or lengthy text descriptions can be referenced to another page or footnote if necessary. If text descriptions refer to more than one box in a flowchart, a dotted line can be drawn around the steps that the text describes or refers to.

The DFDs and workflow diagrams are iteratively developed, often as the data sources and processing needs are being considered. Such iterative development leverages the informative graphical nature of the diagrams to help study teams see what may be missing or to spot redundancy or opportunities for automation. DFDs and workflow diagrams are used throughout the project to communicate how data are acquired or processed, for example, to new team members, to institutional leadership, or to auditors. Finally, after the study is completed, the DFDs and workflow diagrams serve as an efficient communication of data collection and processing and can be shared with the manuscript as supplemental material or shared with data to orient data users to the data they have received.

Summary

This chapter introduced graphic depiction of processes and data flows. Three diagram types were introduced and contrasted using an example of a material testing study. The symbols and conventions for each were presented.

Exercises

1. Which of the following is a reason why diagrams of processes are important?
   a. Information is conveyed faster through images than through narrative text
   b. Process diagrams facilitate visualization of the whole process

c. Crating the diagrams often identifies areas where additional specificity is needed
d. All of the above

Correct answer: (d) All of the above, all choices provide reasons why diagrams of processes are important.

2. Workflow diagrams and DFD depict completely different aspects of processes. True or false?

   Correct answer: False, high-level DFD depict context; whereas workflow diagrams also do so but to a much lesser extent. Workflow diagrams show sequence of steps and roles, whereas DFDs do not.

3. All DFDs are drawn at the same level of detail. True or false?

   Correct answer: False, DFDs can depict different levels of detail. The level of detail within any one diagram should be consistent.

4. Which of the following symbols would be used to depict a study database on a DFD?
   a. Rectangle
   b. Open-ended rectangle
   c. Double-sided rectangle
   d. Circle

   Correct answer: (b) Open-ended rectangle.

5. Which of the following symbols would be used to depict a data transformation on a DFD?
   a. Rectangle
   b. Open-ended rectangle
   c. Double-sided rectangle
   d. Circle

   Correct answer: (d) Circle.

6. Which of the following is the main aspect of a process depicted on a context diagram?
   a. Interfaces between the study and the outside world
   b. Interfaces between data systems used on the study
   c. Data transformations between entities and data stores
   d. Data transformations between entities and other entities

   Correct answer: (a) Interfaces between the study and the outside world.

7. Which of the following is the main aspect of a process depicted on a workflow diagram?
   a. Data transformations between entities and data stores
   b. Process steps and their sequence
   c. Data transformations and their order
   d. Process steps and data transformations

   Correct answer: (b) Process steps and their sequence.
8. Which of the following symbols would be used to depict the following statement on a DFD? "Data collected by the field engineer are entered into the study database and the records are subsequently linked with existing data collected at that location."
   a. Rectangle and two circles
   b. Rectangle, open-ended rectangle, and two circles
   c. Three circles
   d. Two circles and a cylinder

   Correct answer: (b) Rectangle, open-ended rectangle, and two circles. For the stated flow, the data must have an origin, so a rectangle is needed to represent the engineer. An open-ended rectangle represents the study database into which data are entered. Data entry and record linkage are operations performed on data and are depicted by circles.
9. Read the following text section. Select the diagram that best represents the text.

   A large international orthopedic trial is being conducted. The case report form (CRF) data are collected in an online electronic data capture system. In addition, the endpoint of the trial is confirmed secondary fractures. The X-ray and magnetic resonance imaging (MRI) images and associated imaging reports are sent to a central reading facility, where the staff read and interpret the images and associated reports. The reading is recorded on a form and sent to the data center for entry and processing. The study uses a central lab. Samples are sent to the lab where they are analyzed. The lab results are sent to the sites. The lab data are also electronically transferred to the data center once a week from the central lab.

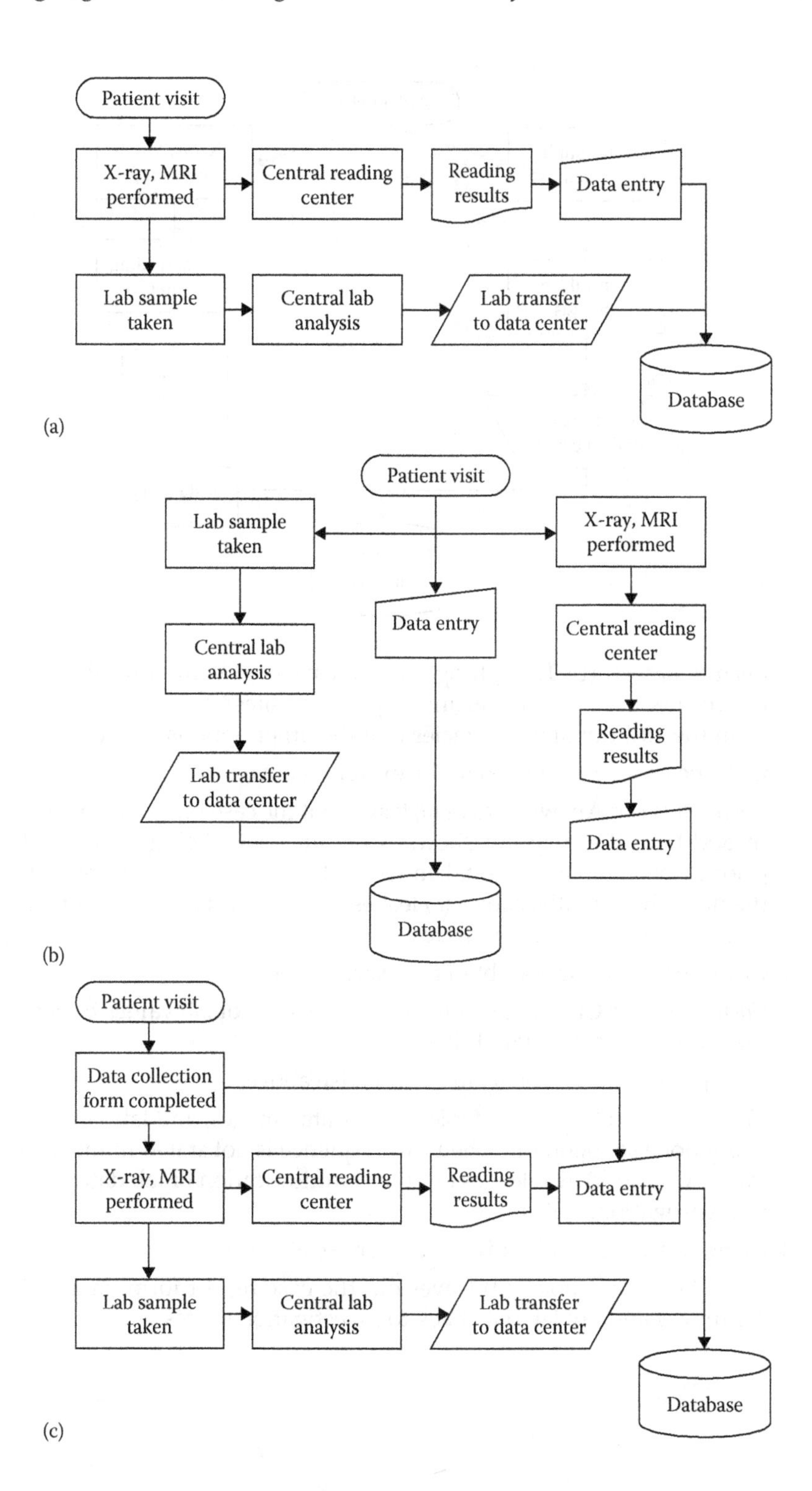
Patient visit
X-ray, MRI performed
Central reading center
Reading results
Data entry
Lab sample taken
Central lab analysis
Lab transfer to data center
Database
(a)
Patient visit
Lab sample taken
X-ray, MRI performed
Data entry
Central lab analysis
Central reading center
Reading results
Lab transfer to data center
Data entry
Database
(b)
Patient visit
Data collection form completed
X-ray, MRI performed
Central reading center
Reading results
Data entry
Lab sample taken
Central lab analysis
Lab transfer to data center
Database
(c)

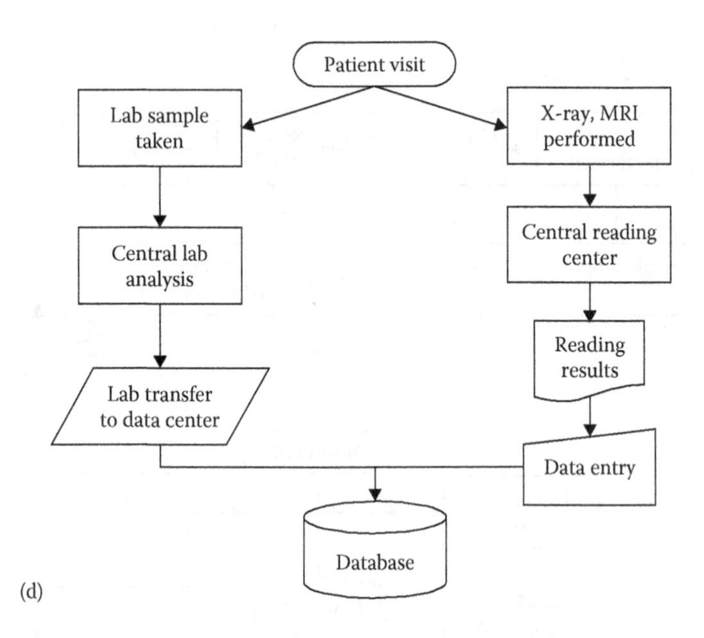

(d)

Correct answer: (b) Though option (b) leaves out communication of lab results back to clinical investigational sites, it does not suffer from the additional inaccuracies that the other options include.

10. Critique answer option (a) in the exercise above.

    Correct answer: Answer choice (a) leaves out the case report form data. In addition, the diagram shows X-ray and MRI being performed prior to lab samples being taken when this sequence is not stated in the narrative. Finally, choice (a) leaves out the communication of lab data back to investigational sites.

11. Critique answer option (b) in the exercise above.

    Correct answer: Choice (b) omits communication of lab values back to the clinical investigational sites.

12. Critique answer option (c) in the exercise above.

    Correct answer: Choice (c) depicts CRF data, images, and lab samples being done sequentially when this sequence is not stated in the narrative. Choice (c) also leaves out the communication of lab data back to investigational sites.

13. Critique answer choice (d) in the exercise above.

    Correct answer: Choice (d) leaves out the case report form data and communication of lab data back to investigational sites.

14. Which diagram would you use to show the interfaces between multiple academic groups on a study?
    a. Context diagram
    b. Detailed DFD
    c. Workflow diagram
    d. None of the above

    Correct answer: (a) Context diagram.
15. Which diagram would you use to show the steps in the data collection process in a grant application?
    a. Context diagram
    b. Detailed DFD
    c. Workflow diagram
    d. None of the above

    Correct answer: (c) Workflow diagram.

References

Wickens CD and Hollands JG. *Engineering Psychology and Human Performance.* 3rd ed. Upper Saddle River, NJ: Prentice Hall, 1999.

Edward Yourdon. Just Enough Structured Analysis Chapter 9, Available free at http://yourdon.com/strucanalysis/wiki/, 1989. Accessed December 1, 2015.

ISO/ANSI 5807 Information processing—Documentation symbols and conventions for data, program and system flowcharts, program network charts and system resources charts, 1985.

Juran Joseph M and Gryna Frank M. (eds.) *Juran's Quality Control Handbook.* New York: McGraw-Hill, 1988.

12

Selecting Software for Collecting and Managing Data

Introduction

Previous chapters have covered getting data into electronic format (Chapter 8), data structures (Chapter 9), and data processing (Chapter 10). Almost always, computers and software are used for these tasks. Software selection for research data collection and processing is difficult, because it relies on decisions about the type of data and processing required and involves understanding the necessary software functionality, knowledge of what software is available, and information about the advantages and disadvantages of different approaches. Further, for a researcher, the ultimate decision is heavily influenced by the individuals available and their skills, institutionally supported software, and cost. This chapter categorizes common software approaches for research data collection and management, describes advantages and disadvantages of each, and provides a framework to help researchers optimally pair available software and human processes in the management of data for research.

Much of what is written about software selection pertains to information and requirements gathering to support decision making for selection of large, enterprise-wide, institutional software products. The focus here is on selection of software available to investigators and research teams. Such software often includes institutionally supported systems specialized for research data management and research documentation in various disciplines, statistical software packages, institutionally licensed and low-cost relational or pseudorelational database management systems, and spreadsheets.

Topics

- Automation and research data management
- Data collection and processing tasks for which computers are often used

- Types of computer systems used for data processing and storage
- Software selection

Automation

One of the goals of this text is to lay plain data management methods and processes so that researchers have the knowledge to choose appropriate data management approaches and tools and articulate them in research plans and budgets. The task tables in Chapters 6 through 10 enumerate the high-level process steps for measurement and recording (Chapter 6), documentation (Chapter 7), getting data into electronic format (Chapter 8), data storage (Chapter 9), and data processing (Chapter 10). These tables list the design and transactional tasks in each area. Task lists, however, do not directly translate into the functionality needed in software. The gap is filled by decisions about which or to what extent each of the tasks is performed by a human or a computer and the processes by which those tasks will be performed. For example, if a computer system utilizes a control file to map data from a data transfer file to the data system, once the control file is written, the loading of data is partially automated. Whereas if a computer system does not have this *functionality,* a computer program has to be written to load data, and the data manager must manually check job logs for *exceptions*. If there is a lot of data loading, a system that supports a control file approach and generates automatic exception reports is desirable, whereas if only one data load is planned, that level of automation is not required. Performing all or part of a set of tasks with a machine rather than through human effort is called *automation* (NRC 1998).

Computers are commonly used for data processing tasks. Although the computer performs the logical (and physical at the machine level) manipulations of the data, interpretation and thinking remain the role of the human. Humans can perform both physical and mental tasks, whereas machines can only perform logical and physical tasks (Figure 12.1). Machines, even computers, cannot think like a human; they can only mimic human decision making by executing programmed logic. In data processing, the computer performs the logical and physical manipulations on the data values, and interpretation and thinking remain the role of the human. In the middle of these two are decisions. Some decisions can be reduced to algorithms (logic manipulations of data values), whereas others cannot. Machines perform decision tasks by executing logic, a set of conditions that a machine can recognize and a set of instructions that a machine can carry out under the conditions. In writing the logic to be executed, a human performs the important task of translating and often reducing human thought into unambiguous step by step instructions that a computer can carry out. Decisions not conducive to computation are those that depend on data or context of which the system is unaware such as human preference

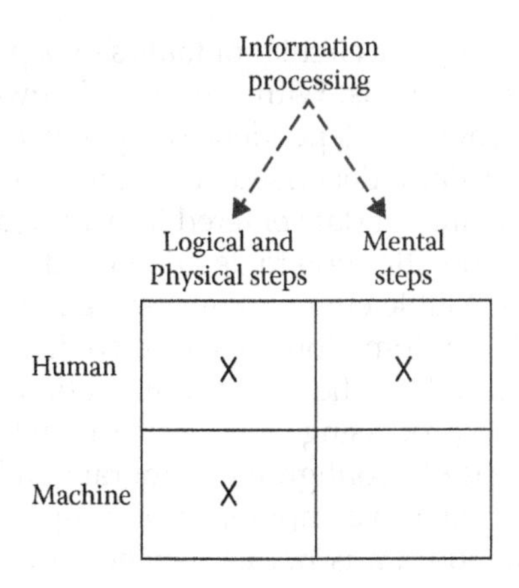

FIGURE 12.1
Physical and mental process steps.

A task should be performed by a human if it is:	A task should be automated if it is:
• Better performed by humans • Expensive to automate • Changes frequently • A fast startup is required	• Better performed by machines • Easily reducible to algorithms • Stable over time • In a situation tolerant of longer startup time • Repetitive and performed with a high frequency

FIGURE 12.2
Factors for choosing to automate a process.

or judgment. Though many decisions can be algorithmitized, doing so for complex decisions increases programming costs (Figure 12.2).

Research settings provide special challenges to automation. For one, research projects are one-off endeavors in which each project is customized to address the questions being asked in the study. As such, individual research studies in general are not conducive to automation. However, very large studies with a high volume of data may be, groups of studies conducted by a particular lab may be, as may certain operations that are conducted in many different types of studies. Collection and management of data from lab analyzers are examples of the latter. Research settings also present challenges due to availability of resources to configure or program automation and carry out testing to assure that it is working properly.

Automation is not only a question of what tasks are more conducive to computation but in research, in particular, also a question of available resources. Take for example the task of building a deck on a house. Such a structure can be built with simple hand tools such as a saw, tape measure, shovel, hammer, and nails. Building a strong and good-looking deck using these tools requires a measure of skill, and the quality is dependent on the skill of the carpenter. The same deck could also be built with the assistance of power and other specialized tools, for example, a circular or table saw, a posthole digger, a drill, nail gun, bolts, screws, and nails. With specialized power tools, many of the manual tasks are aided, and the dependence of quality on individual skill and attention to detail is partially decreased. The same applies in data processing. For example, the quality of data entered into a spreadsheet is completely dependent on the skill and attention to detail of the data enterer, whereas the quality of data that are double entered into a system with such functionality is less so. As the level of automation increases, so does the consistency with which the task is performed. As the level of automation increases, the reliance on skill required for data processing tasks decreases, whereas the necessity of software-specific expertise to configure or program and test the automation increases. For example, where a computer system supports double data entry, the quality of the entered data is not as dependent on the individual data entry operator as it would be in a single entry situation.

In research settings, the available resources may not be completely within an investigator's control. For example, an institution may provide software for data collection and management or access to individuals for data management. When these resources are subsidized by the institution, using other software or people will be more costly to the study budget. In addition, using software other than that provided by the institution may require special approval or may require the investigator to cover the cost of hardware, software, software installation, testing, maintenance, training and use. Because of the inverse relationship between level of automation and data management skill and the positive correlation between level of automation and software skill needed, availability of workforce should inform software selection and vice versa.

Mismatches between the level of automation and data management or data processing staff can be problematic. Consider a scenario in which a researcher has a new undergraduate research assistant and plans to have him or her enter the data into a spreadsheet. There are no individuals with data management experience or training available to the study. The undergraduate research assistant will have to create and document the spreadsheet and data entry procedures and have to enter the data. In this common scenario, a low-skill level individual has been paired with software containing little automation. Little automation means few constraints on the data structure, formatting decisions, procedures for entry and decisions about how to handle aberrant values. The research assistant is not likely aware of the type of errors to expect nor knowledgeable in the design of procedural controls to make up for the lack of automation. Further, with no experienced data management personnel available, the researcher is the only person in the scenario to train the research assistant in data management and processing and to review and provide oversight for the work.

In contrast to the above scenario with one in which a spreadsheet is the planned software, but an experienced data manager is available to the researcher. In this new scenario, the data manager formats the spreadsheet to hold the data collected for the study and reviews the referential integrity between data on various tabs with the investigator. The data manager drafts a data dictionary for the spreadsheet as well as data entry and review guidelines. He or she has designed some procedural controls such as daily backups between version comparisons, manual review of the data entry by a colleague, and several consistency checks that he or she can program into the spreadsheet. In the former case, the research assistant is unprepared to create the documentation and procedural controls that will increase the consistency of the data processing using a spreadsheet, whereas in the latter scenario what the spreadsheet lacks in technical control has been in large part mitigated by procedural controls and independent review. Although this scenario can provide adequate data quality, the cost of setting up the procedural and technical controls to make up for functionality not present in the spreadsheet, and the manual work is often higher than the cost of central support services such as an institutional data collection system and data management resources that may have on the surface seemed more costly.

Consider still a third scenario in which the researcher has an institutional data system available to the study. A centralized support group works with the investigator to draft a data dictionary, create data entry screens, write data entry instructions, and build needed referential integrity in the data. The system only supports single data entry, so the support group also works with the investigator to craft consistency checks and checks for missing data that will run during data entry. The investigator hires an undergraduate research assistant to enter data, and the central support group trains the research assistant on data entry for the study. The central support group also identifies a concurrent study and suggests a reciprocal agreement in which the two research assistants perform a manual review of a representative sample of the entered data for each other. In this scenario, the institutional system provided some technical controls. Procedural controls such as data entry instructions and independent review were added to assure consistency. In addition, the investigator was able to leverage the system and the available data management experience to support appropriate use of lower skill level individuals for data entry.

Both technical and procedural controls have been mentioned. A *technical control* is a part of a process that is enforced by software or performed by the software, which effectively assures that the automated part of the process is performed in the same way each time and at a known level of quality. The functionality of an audit trail is an example of a technical control. An *audit trail* in a computer system tracks creation of new data values, changes to existing data values, and deletion of data values, and in some cases, even tracks each time and by whom data are accessed or viewed. Audit trails store the action taken (creation, update, delete, and view), the user account initiating the action, and the date and time of the action, and sometimes prompts the user to record a reason for the action. The purpose of an audit trail is that all changes and sometimes access to data are recorded. As such, audit trail information is not editable by system users; thus, it is a complete and faithful record of all changes to data. This functionality of an audit trail is a technical control because the audit trail forces the tracking of actions taken with respect to data in a way that guarantees all actions are recorded. However, like most technical controls, audit trail functionality depends on human and organizational actions such as validating the computer system to test that the audit trail is not penetrable, procedures for access to the system such that the integrity of the user ID is maintained, and separation of duties such that technical actions to override or otherwise penetrate an audit trail require collusion of two or more individuals and periodic audits to assure that these procedures are followed. Although the audit trail functionality enforces the workflow, these other human actions are required to assure that the technical control is implemented and maintained appropriately. In a similar way, systems that route documents for review and approval enforce workflow and leverage technical controls. All process automation can be considered technical controls.

Procedural controls on the other hand are implemented by humans, and although they may also be rigid, they are not absolute. A standard operating procedure (SOP) is an example of a procedural control. It is a document that specifies the steps of a work process, their sequence and who performs them. Such work procedures depending on organizational preference may describe in detail how a task is to be performed. Written procedures may in fact be very rigid in terms of what is permitted and how a process is to be performed, but the written procedures in themselves cannot guarantee that the work will be performed as stipulated. As such, procedural controls are weaker controls than technical controls. Procedural controls can be implemented in any situation, whereas in some situations, technical controls may not be cost effective, may take too long to develop, or may not be accomplishable with available technology. To assure consistency of the specified process, procedural controls are usually accompanied by some oversight mechanism. For example, data processed under procedural controls are usually subject to independent verification of a sample or to an audit to assess the extent to which the procedures were followed or the quality of the processed data. Although procedural controls are faster to instantiate, they require additional resources in terms of maintenance of the procedures themselves and the actions taken to assure adherence. A third category of control, managerial control, is also a procedural control and will be discussed in the chapter on quality systems (Chapter 13).

There is a trade-off between automation and flexibility (Figure 12.3). Where technical controls are in place, if a different process is needed, the software or software configuration will need to be changed, and the change will need to be tested and implemented. Changes take time and resources. There are two types of automation called hard and soft automation. *Hard automation* is that which is the more difficult to change, usually requiring computer

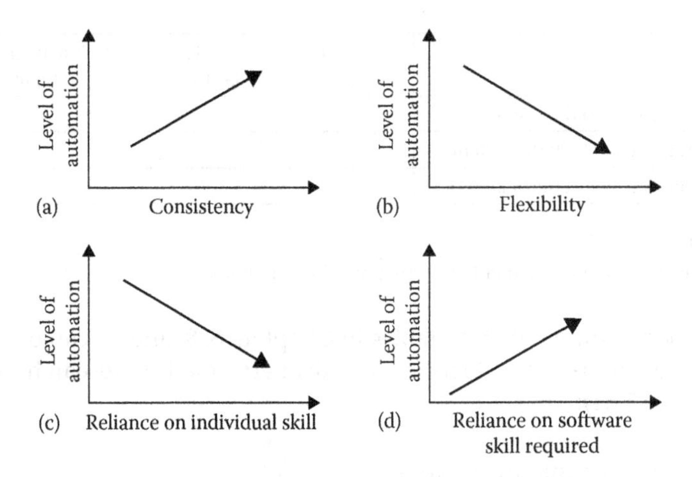

FIGURE 12.3
Relationships between (a) level of automation and consistency, (b) level of automation and flexibility, (c) level of automation and reliance on individual skill, and (d) level of automation and reliance on software skill.

programming to be altered. Changing hard automation takes longer than changing soft automation. *Soft automation* refers to automation that is configurable through the software without requiring changes to the underlying code of the software itself. For example, some data entry systems allow the researcher or data manager to label each data field as required or not. In some systems, the required fields cannot be passed by a data enterer until they contain a valid value. Other systems display a pop-up window requesting the data enterer to confirm that they intend to leave the field blank. Allowing the researcher or data manager to label fields in this way is an example of soft automation because the change of designation (required or not) does not require a change to the underlying software, only to parameters that a data manager provides. Such changes do not require updates or reinstallation of the system, thus the designation as *soft*.

Data Collection and Processing Tasks for Which Computers Are Often Used

Computers show their worth in data management by performing some data-related tasks better than humans. Examples include execution of repetitive or algorithmic steps over large amounts of data, storage and organization of large amounts of data, real-time processing, and management of data over

	Task configuration	Task performance	Task tracking and reporting
Getting data into the system			
Operating on data in the system			
Getting data out of the system			

FIGURE 12.4
Components of data management tasks that may be automated.

long periods of time. The task tables in Chapters 6, 8, and 10 can be generalized to three high-level categories of tasks performed by research data systems (Figure 12.4):

1. Getting data into the system
2. Operations performed on data while in the system
3. Getting data out of the system

Getting data into the system was the focus of Chapter 8. Operations performed on data were covered in Chapter 10. Getting data out of the system occurs during system testing, that is, documenting test cases and system validation; during study conduct, that is, for status reporting, work lists, data transfers to collaborators, and interim analyses; and at study close for the analysis, for data sharing, and ultimately for archival. There are multiple approaches, methods, and workflows for getting data into a system, operating on data in the system, and getting data out of a system.

Getting data into a system, operating on data in the system, and getting data out of a system may be automated to varying degrees. Such automation may include (1) programming or configuration of the task in the system, (2) performance of the task itself, and (3) tracking or reporting of the task (Figure 12.4). Take for example manual key entry. Configuration of the tasks within the system consists of creating data entry screens, creating field order on the screen including conditional skip patterns, setting up workflow for the type of manual key entry desired, creating any on-screen checks that run during data entry, and creating any instructions to be available through the system to the user during data entry. Data systems automate these configurations or set up tasks in different ways and to varying degrees. For example, some systems require the researcher or data manager to fill out a spreadsheet that specifies the sequence of fields on the screen, and for each field, the data collection format, prompt, discrete valid values for enumerated data elements or a valid range for continuous data elements, and whether the field is required. From this information, the system generates the data entry screen. Other systems require computer code to be written to build the data entry screens, whereas others provide graphical interfaces where desired field types and options are dragged and dropped to build the data entry screen.

Spreadsheets provide little of this functionality, and setting up a spreadsheet for data entry entails creating fields, labeling columns, setting limits on columns, and adding any algorithms. The matrix in Figure 12.4 emphasizes that for each of the three main system functions, there are three corresponding areas that may be automated. Because the automation (supported data flow and workflow required of human users) can vary significantly, a matrix similar to that in Figure 12.4 can help compare different systems according to their level of automation and other workflow. Further, several methods of evaluating usability such a keystroke level modeling and related techniques can quantify performance of various systems in these areas so that the researcher can compare, for example, the number of keystrokes and other human actions required to set up data entry screens or data transfers within the systems considered.

Software packages are built for one or more approaches to getting data into a system. For example, recognition-based systems may handle both character and mark recognition. Other software packages focus on speech or handwriting recognition, whereas others handle different types of key entry. Each of these systems is automating to some degree the performance of the task of getting data into a system in a different way. The workflow differs for each of these different approaches to getting data into electronic format, as does the optimal separation of tasks performed by human and machine. In recognition-based methods, the act of transforming data from spoken or written word is automated, whereas in key entry systems, the transformation is done by the human typing on a keyboard while the workflow of identifying discrepancies between the first and the second entry or in the first entry alone and presenting them for resolution is automated. A device such as an electronic blood pressure cuff when interfaced to a computer automates measurement of the physical quantity as well as recording the result in electronic format. Automation comes at a price; a human must configure the systems and test them; the cost of this might not outweigh the savings from automation of the task itself. Chapter 13 covers calculating the cost of data management tasks, and the method described there can be used for comparing the cost of such configuration to the cost saved through automation of the task itself.

Types of Computer Systems Used for Data Processing and Storage

There are four basic types of systems that are used to collect and manage research data, spreadsheets, databases, statistical analysis packages, and systems specialized for collecting and managing research data. Spreadsheets are readily available to most researchers; they are often already installed on computers as part of common office software packages, and most researchers have some experience with using them. Spreadsheets, however, do not have the technical controls to enforce data integrity, an audit trail, discrepancy tracking, or facilities for specialized data entry methods. Data entry screens and discrepancy identification rules and the aforementioned technical controls can be programmed, but each has to be built. Spreadsheets have the added advantage of being able to start work immediately, unless the researcher needs any of the aforementioned technical controls developed.

Like spreadsheets, databases also organize data in logical rows and columns; however, unlike spreadsheets, databases have the ability to enforce integrity between data, for example, that one or more columns in a table must be unique or must not be blank, or that a column in one table should match up with the column in a second table. Some database systems have facilities for creating data entry screens, and for creating, storing, and triggering rules like those that identify discrepancies or transform data. Some database systems also have functionality that tracks changes to data. Unlike spreadsheets, in some database systems, the tables that store the data are created by programming, whereas in others, their creation is facilitated in some way, for example, a drag-and-drop user interface.

Some statistical analysis packages provide spreadsheet-like interfaces for data entry, whereas others provide facilities to create data entry screens. Statistical analysis packages also offer the ability to program rules against the data and to undertake a large variety of data transformations. However, like spreadsheets and databases, they do not support specialized data acquisition such as modes of double entry, optical character recognition (OCR), optical mark recognition (OMR), speech recognition, interfaces with devices or tracking data acquisition and cleaning.

In almost any discipline, specialized data systems exist for data collection and management. For example, electronic lab notebook systems are available for labs, and multiple data systems exist for clinical trials. These specialized systems come with technical controls that automate or enforce practices in data collection and management. An example of one of these technical controls is attaching a date and time and the user identification of the account making the entry to each entry. The system does this automatically, and it cannot be altered by the person entering the data. This functionality enforces the principle of attribution. The user identifier of the account through which the data were entered remain inextricably associated with the entered data. Through association of the date, time, and user identifier, this functionality also supports contemporaneity. When system users know that date and time stamps are automatically applied and that late entries can easily be detected, they are less likely to enter data after the fact. Technical controls, especially those that automate parts of common research processes, make it easy for researchers and research teams to follow minimum standards and best practices. This functionality, however, comes at a cost. The software vendor has to undertake development, testing, and maintenance, and systems with more functionality are more complex and more expensive to maintain. From the researcher's perspective, where this functionality is needed, the researcher has to use procedural controls, develop test, and maintain their own technical controls, use an institutionally supported system that has them, or pay for a system that has them in place.

To summarize the types of systems available, they include spreadsheets, relational database systems, statistical analysis packages that support some data entry features, and open-source or vended software specialized for research data collection and management. Many of the available software products across these categories of systems have a way to perform the basic functions of data collection and management. The products are differentiated by how each of the functions is performed, by the support for different ways of getting data into electronic format, and by the extent to which technical controls come with the system versus have to be developed by the research team or instituted as procedural controls where technical controls are not available and not feasible for the research team to develop on their own.

Software Selection

Whether a researcher is choosing from available institutional software or selecting a system from those available as open source or through commercial outlets, software selection is started by listing the tasks that the system will support and the desired level of automation. Often, the selected system will not meet all of the needs. After a system is selected, the list of desired tasks and automation should be re-compared to the list of actual tasks and automation supported by the software, and decisions made about how to fill the gaps, for example, with custom computer programming or with manual operations and procedural controls.

Consider a study of a random sample of households within five neighborhoods. The researcher is getting household addresses from tax records and plans to mail a 30-item survey, phone those households that did not return the survey within 2 weeks, and then send teams of interviewers door to door for those households not returning a mail survey and who could not be contacted after three phone attempts. If after three visits to the house there has been no contact, the household will be marked as unable to contact. The researcher needs a system to track the 300 randomly sampled households through the survey process and a system in which to enter data.

Tracking tasks include:

- Importing file of the households in the sample (address, owner, and phone number)
- Tracking survey receipt and result of each phone and visit attempt
- Print a daily work list of phone calls needed and household visits needed

Data entry tasks include:

- Association of data to a household
- Single or double entry (double preferred by the researcher)
- Ability to track all changes made to data
- Ability to detect missing values and values that are not valid

At the university, the researcher has several software options available to him or her including a spreadsheet and an institutionally supported web-based data entry system. The two systems available to the researcher have some significant differences.

If he or she used the spreadsheet, he or she would have to set it up for tracking and for entry of the 30 data items. The spreadsheet in the absence of some additional programming only supports single data entry. Further, the spreadsheet can enforce limits on enumerated fields through drop-down lists and valid ranges on numerical data, but invalid data cannot be entered; so the valid values have to be inclusive of any physically possible response. The spreadsheet cannot by itself track changes to data values, so all changes would need to be tracked on the paper survey forms in writing.

If the researcher used the web-based data entry system, he or she would need to complete a spreadsheet of information about each data field and upload it to create the data entry screens. The web-based system will support double data entry where the second entry operator resolves any discrepancies between the first and the second entry. The web-based system has the capability to add programmed checks for missing, out-of-range, and logically implausible data, and there is a central group that helps research teams configure the data discrepancy identification rules in the system. These checks have to be created by a programmer with specialized knowledge of the system. The web-based system has an audit trail and automatically tracks changes to data values. In addition, each data discrepancy identified is given a unique identifier, and the identifier is associated with the discrepant data value and the terminal resolution (overridden, confirmed as is, or data change). The researcher could also use one of the many commercially available inexpensive database packages; however, use of these would require specialized programming not readily available to the researcher to set up data entry screens, build an audit trail and workflow for double entry, track household disposition, and create rules to check the data for missing values or discrepancies. There are no computer programming skills available to the investigator on his or her current research team, so custom programming using the off-the-shelf database packages is removed early from consideration. In addition, the volume of tracking (300 households tracked through a sequential process of multiple attempted contacts) and the volume of data (300 forms with 30 data fields per form) are not large enough to require significant automation.

The researcher is considering using part of the research budget to support 10 hours a week of a data manager and 20 hours per week of a research coordinator. In addition, each student completing a research operations practicum commits 120 hours (basically 10 hours a week) for one semester for course credit. Chapter 14 will cover the methodology to estimate the resources necessary for data collection and processing. For now, considering the human resources available to the project, the students will perform the telephone calls and the home visits (in pairs of two for safety) and will enter the data. The research coordinator is responsible for compliance with the human subjects protection requirements such as submission of study materials to the institutional review board for approval. The research coordinator will also serve as the project manager and will train and oversee the students undertaking the telephone calls and home visits. The data manager is responsible for designing the data structure, for setting up the data system, and for designing and implementing the data discrepancy rules, but this particular data manager is not a programmer.

Given these resources and the available budget (25% of a data manager and 50% of a research coordinator, plus free use of the institutionally licensed software and low-cost central programming support), the researcher decides that the $5500 cost to set up the web-based system for data entry of the survey including up to 20 programmed multivariate data discrepancy checks is well worth it and decides to use the web-based system for data entry of the survey form. The web-based system, however, does not have the functionality to create work ques or easily track the mailing, calling, and visiting necessary to collect completed survey forms. The researcher asks the data manager to set up a tracking spreadsheet and to train the research coordinator and students on use of the spreadsheet for tracking. The data manager does so with one row per randomized household that contains the contact information, one column for the initial mailing (done in four waves of 75 households each), a date column and comment column for each of the three phone calls, and a date column and comment column for each home visit. The date column is to be populated with the date of the contact attempt and the comment column populated with any informative specifics about the contact attempt. The data manager also will work with the central support for the web-based system to set up the entry screens and data discrepancy checks, to test them, to train the students on their use, and to assign and oversee data entry and discrepancy resolution.

TABLE 12.1

Data Management Tasks Supported by Data Collection and Processing Systems

Task	Level of Automation
Storage of data element-level metadata	
Supports manual entry or import of data element metadata	
Supports management of the data element life cycle, for example, draft, active, and deprecated statuses for data elements	
Tracks changes in data element metadata	
Stores mapping of data elements to relevant common data models	
Study-specific configuration	
Supports study-specific data structures and referential integrity	
Supports study-specific screens and workflow	
Supports use of study- or data-specific devices for data entry (smart phone, tablet, PC, and others)	
Supports off-line device use, i.e., collection of data when the device is not connected to the internet and automatic uploading of data when a connection is established	
Supports study-specific account management	
Creation of data entry screens	
Facilitates selection, reuse, and editing of existing screens	
Supports development of new data entry screens	
Supports custom layout of fields on screen, for example, nesting fields or locating fields beside other fields so that the data entry screen co-locates related data and can have any layout that a paper form could have	
Supports free text, user-defined semistructured text, type-ahead text entry, select all check box, select one check box, drop down list with user-defined item order, radio button, user-defined image map, image and wave form annotation, and visual analog scale entry	
Supports conditional skip patterns on data entry screens	
Supports conditionally generated forms at run time, i.e., generates a new form for a user to complete based on the response to previously entered data	
Supports univariate on-screen checks, for example, missing, valid value, and range	
Supports multivariate including cross-form on-screen checks	
Data entry	
Supports choice by screen of single versus double data entry	
Supports double blind or interactive data entry	
Supports third-party compare for blind double entry	
Supports manual visual verification of single entry (flagging or tracking errors)	
Supports on-screen checks for all entry types	
Supports tracking of data discrepancies identified via on-screen checks	

(*Continued*)

TABLE 12.1 (*Continued*)

Data Management Tasks Supported by Data Collection and Processing Systems

Task	Level of Automation
Import of electronic data	
Supports importing (also called loading) of electronic data from delimited files, spreadsheets, and other file formats	
Supports cumulative and incremental loading	
Supports real time, for example, messaging and direct interface data acquisition as well as by-file loading	
Generates report of data failing to load and user-defined load exceptions	
Maintains audit trail for all loaded data, including user-defined load parameters	
Supports user updates to loaded data and manual record linkage corrections	
Supports user-defined referential integrity between loaded and other data	
Supports loading and integration of narrative, image, and waveform data files	
Character and voice recognition	
Supports voice recognition of audio files	
Supports mark (OMR) and character (OCR; optical character and intelligent character recognition, ICR) recognition	
Supports user-defined recognition confidence	
Supports visual verification and update of fields below confidence	
Supports creation and transmission of user-defined data clarification forms during visual verification	
Supports user-defined referential integrity between recognition and other data	
Data cleaning	
Supports creation and management of rules to identify data discrepancies	
Supports manual identification of data discrepancies	
Supports discrepancy identification, communication, and resolution workflow	
Tracks data discrepancy life cycle including terminal resolution	
Data transformation	
Coding	
Supports auto-encoding with manual coding of uncoded values	
Supports versioning of controlled terminologies	
Supports manual discrepancy identification during manual coding	

(*Continued*)

TABLE 12.1 (*Continued*)

Data Management Tasks Supported by Data Collection and Processing Systems

Task	Level of Automation
Algorithmic value change	
Supports user-defined imputation, reformatting, mapping, and calculation	
Supports user-defined anonymization and deidentification	
Tracks algorithmic changes to data	
Supports deterministic and probabilistic record linkage	
Data export	
Stores multiple mappings for data sharing	
Stores interim analysis or other data snapshot	
Supports user or other system SQL and open database connectivity (ODBC) access	
Supports preservation of study-related data documentation (document management)	

The survey study example is very realistic in terms of the software available to researchers at low or no cost and in terms of the skill sets available to the researcher. The example illustrates the necessary decisions about the needed level of automation, the cost of such specialized automation, and the lower availability of those very specialized technical skills. In addition, the scenario is quite realistic in which no one on the research team has computer programming experience that would enable the researcher or data manager to oversee one or more programmers developing a custom system. Using staff that the researcher is not qualified to oversee is risky. The example is also very realistic in the depiction of the interplay between the human skills available and matching those with available software. In general, the more automation available from a system, the more expensive the system. In addition, automation takes more specific skills and knowledge to appropriately leverage and set up. Thus, while on the surface, a researcher might want the most automation possible, high levels of automation may not be available or cost effective.

The task tables in Chapters 6, 8, and 10 list high-level tasks for configuration, task performance, and tracking of common data-related tasks. The task tables are very high level. To use them in software selection, additional detail is needed. For example, in Table 10.2, the task *Specifying and documenting the data cleaning procedure* means the workflow involved in data cleaning. For a particular study, a researcher may intend for the discrepancies to be flagged but not corrected, or a researcher may intend that each discrepancy be communicated to the person collecting the data for resolution and the database updated based on the resolution. The steps in the workflow, whether they are performed by a human or a computer, and when and how they are done will all be outlined in the workflow. In software selection, the intended workflow needs to be compared to workflows supported by each candidate system. Some systems will be a close match, and others will not. Specifying the intended workflow is necessary for assuring a good match between a system and the intended workflow of a given study. The data collection and management tasks listed in Table 12.1 may aid in this process by prompting discussion about the desired versus available level of automation for the listed tasks that are relevant to a study.

In a software selection process, the high-level tasks are listed; then a scenario or test case is written depicting an expected workflow that the selected system would need to support, so the researcher can ask a vendor or representative for institutionally supported systems to show how the system performs the task to see exactly how a given task is performed in the candidate system. In the data cleaning example above, if a researcher intends for data discrepancies to be communicated back to the data collector for resolution, an example scenario would be: *show how the system identifies a data discrepancy and facilitates communication and resolution*. There are many ways that a system can do this, some of which may or may not be a good match for the study. In the case of institutionally available software, once a researcher understands how the system facilitates the needed tasks, the researcher may adopt the way the software supports the task. The key of course is finding a good match between how software enables a task and the needs of a particular study. When institutions select software that will be broadly used, a matrix is often created with the tasks needed—with workflow specifics—listed down the left-hand column and each candidate system listed across the top (one system per column). Using the matrix, each system is evaluated against each task, and the software selection team can easily see which system is the best match. For a small research study, this approach is too onerous, but the concept of matching workflow for key tasks between software, a research study, and available skills helps select the optimum system.

Summary

This chapter covered the concept of automation with respect to research data management, describing the relationship between automation and technical controls, and how these may be available in some software for research data collection and management. The concept of software functionality as actions or process steps performed or facilitated by software was introduced. Data collection and processing tasks for which computers are often used were discussed, as were types of computer systems used for data processing and storage. Finally, these concepts were woven together to describe the interplay between needed functionality, level of automation, and availability of skilled resources in software selection for research data handling.

Exercises

1. Draw a workflow diagram for the neighborhood survey study in the example given in the chapter.
2. Which of the following are ending terminal statuses on the workflow diagram from number 1?
   a. Survey mailed
   b. Survey received
   c. Third phone contact failed
   d. Unable to contact

   Correct answer: (b) and (d). Depending on where the student started the process, survey mailed may be a beginning terminator. Third phone contact failed is not a terminator because household visits follow the third failed phone contact.
3. Describe the difference between technical and procedural controls.

   Correct answer: Technical controls are the aspects of processes that are hardwired in the system or otherwise enforced. Procedural controls on the other hand are guided by written procedures. There is no enforcement for procedural controls within a computer system. However, procedural controls may be enforced by other procedural controls, for example, random audits to assure that procedures are being followed.
4. A researcher needs to track data forms received from a mail survey. He or she specifies a process in which the mail for the study is opened by the research assistant, and the address as well as the date received are recorded on a coversheet that is affixed to the data collection form. This is an example of which of the following?
   a. Technical control
   b. Procedural control
   c. Managerial control
   d. None of the above

   Correct answer: (b) Procedural control. This is an example of a procedural control because the researcher has specified a process (procedure) to be followed, but there are no guarantees that the process will be followed, for example, the date could be omitted, an address could be lost and not recorded, and so on.
5. In the question above, if the survey were conducted via a web-based system, what technical controls would likely be in place?

Correct answer: If the survey was conducted via the web, the date and intended respondent identifier could be automatically associated with the collected data by the system. In the web case, there is a guarantee via automation that the process will be followed, because it is automated and not subject to human error or noncompliance.

6. For a research study, data are received via data transfer from an external lab as well as collected via the web directly from study participants. List the main functions needed in a data collection and management system.

   Correct answer: This question can be answered at different detail levels. At a high level, the functions (tasks the system needs to support) should fall into the following categories:

   Support the development of web-based forms

   Support the development of on-screen data discrepancy checks

   Store the entered and imported data

   Associate entered data with the correct research participant

   Import external data

   Store external data

   Associate external data with the correct study participant

7. Which of the following should a research team be prepared to handle on their own, that is, not assisted by software, if a spreadsheet is used to enter and store study data?
   a. Set up places to enter data
   b. Manually track changes to data
   c. Check for data discrepancies
   d. All of the above

   Correct answer: (d) All of the above. Spreadsheets do not typically have functionality to facilitate creation/definition or formatting of the columns in which to enter data, tracking changes to data, or checking for data discrepancies.

8. What principle is supported by functionality that tracks changes to data?
   a. Originality
   b. Traceability
   c. Accuracy
   d. none of the above

   Correct answer: (b) Traceability. Tracking changes to data facilitates the ability to trace a value from the analysis back through all changes to it to its original capture.

9. What principle is supported by functionality that associates the account making an entry to the entry?
   a. Originality
   b. Traceability
   c. Attribution
   d. none of the above

   Correct answer: (c) Attribution. Associating the account making the entry to the entered data value links or attributes the entered value to that account.
10. True or False: Using a spreadsheet is always the least expensive option for managing data.

    Correct answer: False. While the spreadsheet may be the least costly software, the total cost of the time and effort for the data processing tasks and addition of any needed technical or procedural controls often trip the balance, especially when data volume is high.

Reference

Automation is the independent accomplishment of a function by a device or system that was formerly carried out by a human. (Source: National Research Council (NRC), 1998; Parasuraman & Riley, 1997).

13

The Data Management Quality System

Introduction

In any research project, no matter how many people are involved, the researcher is ultimately responsible for the ability of the data to support research conclusions. This book presents methods that produce reliable and accurate data; however, the infrastructure and environment in which these methods are applied also have bearing on the outcome. Quality management theory and practice also take this holistic perspective. This chapter combines theory and approaches in quality management and applies them specifically to the challenge of managing data in research settings to define a data management quality system.

The purpose of such a data management quality system is to assure that data are collected and processed consistently and that data are capable of supporting research conclusions. This chapter presents the major factors that impact data quality and prepares the reader to control the immediate environment in which his or her project occurs using a quality management system approach.

Topics

- Quality management systems
- Defining quality
- Frameworks for managing quality
- A quality management system for research data management

Quality Management Systems

Mahatma Gandhi is reported to have said that "Happiness is when what you think, what you say, and what you do are in harmony." Integrity has similarly been defined as one's thoughts and words matching their actions. This concept, the matching of intent, statement, and actions, is a foundational principle of assuring quality. It gives others confidence when an organization's or an individual's intention is to do the right thing, and when the organization or individual documents that intent and makes provisions for the positive intent to be carried out. In organizations and research projects, this triad often takes the form of policy (stating intent of what is to be done), procedure (documenting the process by which it is done), and the output (the actual actions performed and the resulting data). One component of an audit of research involves traversing these relationships: (1) comparing a research plan, for example, as stated in a grant application or an organizational policy with applicable regulations, (2) comparing procedures to the research plan or policy, and (3) comparing what was actually done, for example as evidenced by the data, documentation, or other artifacts to the stated procedures. *Integrity of research operations* can be demonstrated by the matching of these in such an audit.

Institutions have varying levels of *policy* and procedural infrastructure. Depending on available procedural infrastructure at an institution, a researcher may bear the burden of drafting and maintaining documented policy and procedure. These may apply to an entire department, to a research lab, or to a particular project. Project-level documentation is greatly simplified where lab, department, or organizational policy and *procedure* exist. In such a case, project-level documentation relies on references of the existing policy and procedure and covers only that not already covered in lab, department, or organizational policy and procedure.

A second way of demonstrating research integrity is through reproducibility. Science demands that hypotheses are falsifiable and that the results be reproducible (Irizarry et al. 2011).* The ability of science in many areas to demonstrate reproducibility of research is increasingly being questioned (Nature 2014). In 2015 alone, multiple journal retractions cited lack of ability to reproduce the research. Much attention has been given to reproducibility in preclinical research (Ioannidis 2005, Baker 2015, Collins 2014). At the time of this writing, severable notable initiatives either provide for reproduction as a condition of funding (PCORI Methodology solicitation 2014) or actively seek to replicate published results (Nosek 2012).

* Historically the word reproducible has been used to mean that the results of a second and consistently designed and conducted study by independent people would provide results supporting those of the initial study. Today, however the language has shifted. Reproducibility refers to independent people analyzing the same data (the data from the initial study) and obtaining the same results, while replication refers to independent people going out and collecting new data and achieving the same or consistent results.

Replication is the most independent reproduction of research where a study is conducted a new by independent investigators using independent data, methods, equipment, and protocols (Peng 2009).

When some reproduction or replication is attempted and fails, attention turns to traceability in hopes of explaining why different results were obtained. *Traceability* is the ability to traverse all operations performed on data from a data value in a result set, table, listing, or figure back to the *origin of the data* value or vice versa. It is the second component of an audit of research. It requires statement of how data will be handled, that is, each operation performed on data, and evidence that the operation was performed. Examples of such evidence include reference to the data value before the change, the date and time of the change, the data value following the change, the entity performing the change (person, machine, or algorithm), and specifications for operations performed. Particularly for changed data values, some also require a stated reason for the change, but this is also often covered by the specifications for operations performed on data. For example, guidelines for operations done by humans define which values will be affected and how changes will be made; specifications for algorithmic operations document the same. Although traceability is not reproducibility, it answers many questions in the case of failed reproductions and replications. As evidence, demonstrated traceability gives research consumers confidence that data were handled according to plan, that data are free from adulteration, and that nothing untoward occurred. These two concepts, traceability, and alignment of intent, statement, and action are keys in assuring the quality of data on which research conclusions are based.

Defining Quality

The American Society for Quality provides two definitions for quality:

> (1) The characteristics of a product or service that bear on its ability to satisfy stated or implied needs; (2) a product or service free of deficiencies (ASQ entry for quality 2016)

The ASQ glossary (http://asq.org/glossary) follows with perspectives from two thought leaders in quality, Joseph Juran and Philip Crosby, defining quality in two ways, "fitness for use" and "conformance to requirements," respectively. Both of these perspectives are commonly used in practice. Walter Shewhart, W. Edwards Deming, and Genichi Taguchi, also thought leaders in quality management, conceptualize quality instead as "on target with minimum variation," a definition based on a distribution of multiple values rather than on a binary decision about a single product being fit for some use or within some tolerance. "On target with minimum variance" relies on both the *central tendency* and the variability with which the products are produced. This thinking places the emphasis on the process by which products were produced in addition to conformance of individual products.

The three perspectives, the broader *fitness for use*, the more narrow *conformance to requirements*, and the process-oriented *on-target with minimum variation*, may use different words and may be operationalized differently, but in the ideal case, where requirements match the customers' needs and the process through which the product or service is produced or provided is designed and managed to meet the requirements, the three in fact intend the same thing. However, practice often drifts from this ideal case. For example, in cases where requirements are defined in the absence of understanding the customers' needs, or where requirements otherwise do not reflect the customers' needs, *fitness for use* and *conformance to requirements* diverge. Likewise, when processes are designed in the absence of or for different requirements, *on-target with minimum variation* diverges from the customer's requirements. These three things, (1) understanding of the customers' requirements, (2) measuring conformance to requirements, and (3) designing and managing processes capable of consistently achieving the requirements, need not be different (Figure 13.1).

The concept of consistency, another way of saying lack of variation, plays an important role in defining quality. As some data-handling processes in research rely on humans in the design, testing, and actual performance of tasks, assessment of consistency is especially necessary. Although automated systems may drift over time or behave oddly in the presence of unanticipated input, humans are additionally subject to other and often greater sources of variation such as distraction, fatigue, and human error as described in Chapter 6. In either case, man or machine, consistency must be built into the process rather than left to the skill, diligence, and expertise of individuals.

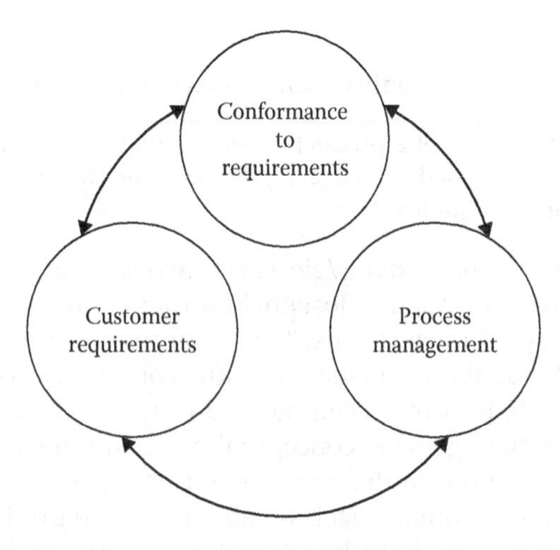

FIGURE 13.1
Different perspectives on quality.

Frameworks for Managing Quality

We learned from cognitive science (Zhang and Norman 1994) that knowledge can exist in the world or in the heads of individuals, referred to as internal versus external representation, respectively. Zhang and Norman showed that human performance improves when knowledge is represented in the external world versus retained and recalled from the minds of individuals. Thus, consistency is improved by leveraging external representation. This can be achieved through technical controls, for example, workflow enforced by computer systems or physical constraints, or job aids supporting procedural controls such as checklists or mnemonics. The consistency of performance over time is increased when knowledge is encoded in an institution's technical controls and procedures rather than maintained in the heads of individuals.

Consistency of performance at an organizational level, whether through man, machine, or both, is the "... with minimum variation" part. The need for managing organizational consistency gave rise to a framework for describing, explaining, and predicting the likelihood of good versus poor organizational performance. The initial version of the capability maturity model (CMM) was published in 1987 by the Software Engineering Institute. Today, the model called the CMM Integration™ or CMMI™ is a five-level framework that describes how organizations across multiple disciplines improve their capability for consistently delivering high-quality products and services. Like the Deming and Shewhart philosophies, the CMMI™ takes a process approach, that is, the focus is on improving the process through which a product or service is produced (Figure 13.2).

The first level in the CMMI™ framework is called *Initial*; processes in organizations at this level are not predictable; the products or services produced are not of consistent quality nor are they consistently produced on time with any regularity. Surprises are common. The lack of consistency is the evidence that processes are poorly controlled and managed. The processes in organizations at the second level, *Managed*, of the model are defined and managed

Maturity Level	Description
5. Optimizing	The focus is on process improvement.
4. Quantitatively managed	Processes are measured and controlled.
3. Defined	Processes are characterized for the organization. Processes are customized for each project as needed. Processes are proactive.
2. Managed	Processes characterized on a project-by-project basis. Processes are often reactive.
1. Initial	Processes are not predictable. Processes are poorly controlled and managed.

FIGURE 13.2
Maturity levels.

on a project-by-project basis. Thus, there is lack of consistency between projects. Processes are reactive to problems when they arise. The third level of the framework is called *Defined*. Processes in organizations at this level are characterized across the organization. Organizational processes are customized for individual projects, but by virtue of defining and managing processes across the organization, consistency of performance across projects is most often the case. Processes at the *Defined* level are proactive. Processes in organizations at the forth level of the framework, *quantitatively managed*, are measured and controlled in addition to being defined. Surprises are few, and consistency and predictability are the norms. Organizations at the fifth and final level of the CMMI framework, *optimizing* focus on improvement. Organizations at the *optimizing* level have achieved predictability and turn their attention to continuing to improve processes. Increasing the maturity of an organization up through the CMMI levels is a move from the consistency or lack thereof depending on the skill, diligence, and knowledge of individuals to achieving a state where the consistency is instead dependent on organizationally defined and managed processes and is thus predictable. Through the work of Sasha Baskarada (2009) called the information quality model–capability maturity model (IQM-CMM), the CMM approach has been applied to organizational information quality (Figure 13.3a).

The organizational maturation journey involves encoding organizational policy and methodological knowledge into organizational processes. This is done through the careful definition and documentation of processes and the design of technical, procedural, and managerial controls (described in Chapter 1) into work processes. Thus, as experienced individuals move on,

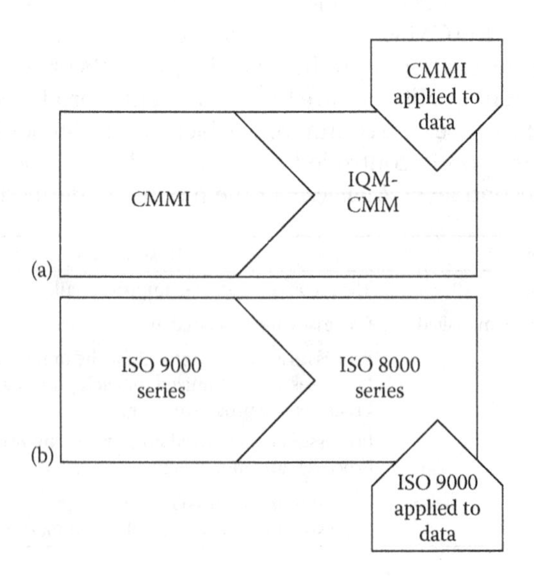

FIGURE 13.3
General standards customized for data quality.

their qualified replacements enter into a stable process and are able to continue the work with consistent results.

The concepts described in CMMI™ model are very applicable to the management of research data. In industry research and development environments, controls such as those described in the CMMI™ model are common place. In small organizations and academic settings, they are less so. Designing and maintaining technical, procedural, and managerial controls require attention and some amount of dedicated resources. Small organizations, companies, and academic research labs alike often struggle to fund or allocate resources to maintain such infrastructure. Although a CMMI™ level 4 or 5 infrastructure would be rare in these settings, the principles of infrastructure to maintain consistency are equally applicable to small organizations and should inform quality systems in small organizations.

The second framework of international importance and relevance to management of research data are the International Organization for Standardization (ISO) Quality Management System (QMS) standards. The ISO 9000 standard articulates seven quality principles and describes the fundamental concepts and principles of quality management (ISO 9000:2015). ISO 9000 defines a *QMS* as "a framework for planning, executing, monitoring, and improving the performance of quality management activities" (ISO 9000:2015). The ISO 9001 standard lays out a path through which an organization can operationalize the quality principles; the path is called a QMS (ISO 9001:2015). Similar to the CMMI™ model, the purpose of the QMS standard, ISO 9001, is to help organizations *consistently provide products and services that meet customer and applicable statutory and regulatory requirements.* The ISO 9000 series of standards defines quality as conformance to customer requirements. At the same time, the standard takes a decidedly process-oriented approach by supporting use of Deming's plan do check act (PCDA) methodology for managing processes and the interaction between processes and does so in a risk-based context. At the time of this writing, there are over 1.2 million organizations certified to ISO 9001. Both the ISO 9000 and 9001 standards emphasize that the QMS needs to reflect the needs of the organization and that QMS components can be integrated into how an organization manages itself rather than some separate set of requirements and procedures. Similar to the CMMI™ approach, full ISO 9000 certification may not be feasible for a small research lab, indeed, management and consistency of processes at the organization level runs counter to the autonomy enjoyed by many academic researchers. However, the QMS principles if incorporated into how a research lab is run and managed will likely be beneficial. The ISO 8000 series of standards applies QMS concepts to data quality (ISO 8000–8 2015; Figure 13.3b).

The CMMI™ and ISO 9000 series of standards have many things in common, including (1) holistic approaches that include organizational infrastructure, leadership, and environment; (2) process-oriented approach to consistency of performance; and (3) inclusion of both prospective assurance and ongoing measurement and control (Figure 13.4). Describing the prevailing

Intent, operational plans, and actions should be aligned.
Organizational environment and infrastructure impact processes and outcomes.
Quality outcomes are achieved through predictable processes.
Both prospective quality assurance and ongoing quality control are necessary.
  There are three types of control: technical, procedural, and managerial.
  Control involves measurement and use of feedback to correct processes.
Traceability is built into processes and evidenced by artifacts and data.
Data quality assurance and control are how data are collected and managed, not something extra or some additional process.

FIGURE 13.4
Key quality management principles relevant to data.

frameworks for quality management should in no way be construed as a recommendation for an individual researcher to pursue certification to CMMI or ISO 9000/1— certification is a major organizational undertaking. However, the principles behind these QMS frameworks can and should inform decisions about quality management in research, and provide investigators (who maintain ultimate authority, responsibility and accountability for research results) a conceptual framework for assuring and achieving their quality.

The principles in Figure 13.4 refer to prospective quality assurance as well as ongoing quality control. With the caveat that the two terms are sometimes used interchangeably, the American Society for Quality (ASQ) provides the following definitions for *quality assurance* (QA) and quality control (QC).

> Quality Assurance is: all the planned and systematic activities implemented within the quality system that can be demonstrated to provide confidence that a product or service will fulfill requirements for quality. (ASQ entry for QA/QC 2016)

Planned or prospective means decided in advance and put in place before a process or a study begins. QC is defined by ASQ as

> the operational techniques and activities used to fulfill requirements for quality. (ASQ entry for QA/QC 2016)

QC activities are usually performed by those involved in the process itself and performed along with the process to periodically or continuously measure quality and take corrective action if necessary.

A Framework for Quality Management of Research Data

Researchers need a way to anticipate, detect, and prevent problems like those described in the data gone awry scenarios in Chapter 1. The principles of QA, QC, and QMSs can be applied in a pragmatic way to anticipate, detect, and prevent data-related problems. QA and QC are not separate from the operations of collecting and managing data; rather, they are the operations of collecting and managing data and as such run throughout a data management plan. As such, QA and QC occur as part of the research process rather than a set of additional activities.

The list in Figure 13.5 is an application of QMS thinking to managing research data. The approach here has been applied in multiple settings from centralized data management groups to research labs and to individual projects. As such, the approach makes key quality management principles (Figure 13.4) accessible to every research project by applying them directly to the collection and management of data for research. Figure 13.5 lists ways in which quality management principles can be operationally applied to collection and management of data and labels each according to whether it primarily serves a QA or control function, the type of control involved, and the evidence produced. Note some items on the list by existing are themselves evidence of the application of quality management principles. Further note that many of the quality management principles are themselves part of the data management plan (recall Figure 4.1).

After a conceptual understanding of the concepts discussed thus far in the chapter, nothing in Figure 13.5 should come as a surprise. The first listed component is organizational policy and interpretation of applicable regulations. Many institutions promote regulatory compliance and mitigate risk by establishing policy around an institutional interpretation of applicable regulations. Often these are only applicable to certain aspects of a research project, for example, human subjects protection, lab safety, animal welfare, environmental protection, or financial management. Researchers should be aware of organizational interpretation of applicable regulation. Likewise, most research-based organizations state some policy for research conduct, documentation, and sharing. Such institutional policy and regulatory interpretation should cover regulations applicable to the research and institutional requirements for research conduct. Existing institutional regulatory interpretation and policy should guide practice and decrease the amount of documentation required by the researcher. Where institutional regulatory interpretation does not exist, as covered in Chapter 1, the researcher must seek out institutional officials responsible for regulatory compliance, identify applicable regulations, and come to a common understanding of things required of the researcher for compliance. Where institutional policy for research conduct, documentation, and sharing does not exist, the researcher must decide his or her own. Documented, that is, written, regulatory interpretation and institutional policy are QA instruments. The presence of such documentation is an evidence that policy-level QA exists.

Institutional policy and regulatory interpretation often specify necessary procedures. For example, a policy for human subject protection might reference institutional procedures for research submission, review, and approval by an ethics committee. Institutional policy on records retention might stipulate that departments establish a discipline-appropriate retention period, a list of documentation to be retained, a process and resources for archival and retention of research documentation. Likewise, an institutional code of

Data management Quality System Component (Necessary Structure)	Assurance	Control Type	Evidence Produced
Quality policy and institutional interpretation of regulations	X		IS
Standard operating procedures (SOPs) and supporting documents	X	P	IS, Specifies
Position descriptions or other job expectations	X	P M	IS
Training for job tasks (role or position-based training)	X	P M	IS, Produces
Management oversight (review and feedback)[a]	X	M	Produces
Computer systems and other technology for data management		T	IS, Produces
Software quality assurance	X	P	Produces
Project management (tacking scope, budget, timeline)[a]		P	IS, Produces
Data quality assessment including internal auditing[a]		P	IS, Produces
Quality system mgt. including independent auditing[a]	X	P	Produces

T, technical control; P, process control; M, management control.
Evidence produced: IS, the item itself is an evidence that it exists; Specifies, the item describes (specifies) the evidence that will be produced during the course of following procedures.
[a]Produces and uses metrics: These metrics are process measures and measures of process outcomes.

FIGURE 13.5
Function of and evidence produced by data management quality system components.

Training of those collecting handling samples and data
Documentation requied for research projects
Procedures for data collection
Procedures for data entry
Data discrepancy identification and resolution procedures
Data transformation and other data handling procedures
Acceptance and integration of externally managed data
Design, programming, and testing of computer systems used in research
Access to research samples and data
Research data quality control
Research database lock and archival
Analysis of research data

FIGURE 13.6
Data-related procedures.

conduct for faculty usually includes research conduct. Departments or more often research labs will establish procedures for sample or data collection and handling that promote compliance with the high-level institutional policy. As part of a research data management quality system, such procedures should cover all operations performed on data from collection to analysis, sharing, and archival, as well as design, programming, and testing of algorithms that perform data manipulations. A list of procedures that are commonly applicable is provided in Figure 13.6.

Procedures are both QA and QC instruments. As a QA instrument, procedures prospectively specify what will be done and the process by which it will be accomplished. As an instrument of control, through describing what will be done and the process by which it will be accomplished, procedures are in themselves a procedural control. The control is accomplished through specification in tandem with assessment of the fidelity with which the procedure is carried out and use of the assessment to improve the process. Further, procedures should be auditable. *Auditable* procedures are those that specify and require artifacts that evidence adherence to the procedure.

Position descriptions, training, and management oversight are the three managerial controls in the framework, and they work together. They must all be present and must support each other or their effectiveness is substantially decreased. Position descriptions usually state the reporting structure (the manager overseeing individuals in the position), job responsibilities or tasks, and minimum qualifications for individuals in the position. The job responsibilities or tasks set the expectations for what individuals in the position are expected to do. Some go further to indicate performance expectations, for example, submission of project reports according to specifications and on time. Minimum qualifications indicate the knowledge and experience that those individuals are expected to enter the position with. When individuals are hired, managers build on the position descriptions and review responsibilities (or tasks) and set performance expectations. At this time, managers also discuss with new hires how their work will be reviewed and what feedback they can expect as they learn the position and organizational policy and procedures. The manager then must follow through and provide this oversight both initially and in an ongoing basis.

In mature disciplines, degree programs exist to prepare graduates with foundational knowledge and relevant skills for careers in the discipline. In this scenario, training at organizations focuses on organizational policy and procedure, that is, how things are accomplished at that specific organization or research lab rather than foundational knowledge and skills; the latter are expected of degree program graduates. Organizational training programs are usually role or position based and focus on organizational policies and procedures. Many organizations maintain training matrices that list the training required for each role or position. Where there are no degree programs supporting a profession, organizational training programs bear the additional burden of training foundational knowledge and relevant skills.

Training, managerial oversight, and position descriptions work together to assure that individuals are qualified for the jobs that they perform and that they perform at the expected level. Training, managerial oversight, and position descriptions are instruments of QA. Training and position descriptions (as used in hiring and evaluation processes) also serve as procedural controls. Training, managerial oversight, and position descriptions, as they are used to assure that qualified individuals perform as expected, are also managerial controls because they are the tools used by the manager to structure and fill positions and to evaluate performance of individuals in positions. Existence of each demonstrates that that component of the quality system is in place. Further, both training and management oversight provide artifacts, for example, training certificates or other documentation of completion, and feedback provided to individuals from managerial oversight or other review that show these components of the quality system are functioning.

Computer systems and other technology for the management of data together are an important component of the quality system. They may provide extensive automation and other technical control that assure, enforce actually, that policy and procedure are followed. In different research settings, more or less technical control may be appropriate. Availability of the needed levels of automation and other technical control is a key aspect of quality management in research data. Where such system functionality is not available, a researcher or lab must supplement the needed controls as procedural controls and ongoing monitoring shows that the procedural controls are effective.

Where computer systems are used for data collection and management, procedures for software QA are required. These are called out specifically and separated from other procedures as a quality system component because these software design, development, and testing activities are often not anticipated by researchers or all together lacking. However, where computer systems are used in research, there is a need to have some procedural controls over the design (or selection in the case of open-source or commercially available products), development (or *configuration*), and testing of these systems. This also applies to algorithms developed by researchers and research teams within computer systems. For example, algorithms to transform, code, or calculate values. These algorithms operate on, change, or otherwise manipulate data and must be specified, programmed, and tested against the original specification or intent to assure that they are working properly. In the case of commercially available or open-source software used to collect and manage data, the system must be tested to assure that it is configured appropriately and working properly. Thus, wherever computer systems are used in research, a procedure for testing that the system works properly over the range of anticipated data should be in place. Those designing and developing such systems should have procedures that cover these activities. In some regulated environments and in federal contracts, the requirements for software QA are much more stringent. Further, testing often can not anticipate or test all possible conditions or nuances of incoming data, thus, monitoring that the system works properly once in production use is a best practice.

Project management is also a necessary component of a research data management quality system. Project management provides a structure in which study operations are planned, tracked, and managed. Project management usually includes a list of project tasks that can be high level or detailed; a timeline associated with those tasks, effort necessary to accomplish the tasks, other resources needed; and the cost. An individual in the role of a project manager tracks the resource use, cost, and timeliness of project tasks. An individual in the role of project manager usually holds meetings with research team members and uses the progress against the plan and metrics to troubleshoot and solve challenges that arise to bring the project's cost and time trajectory back into control. Where this is not possible, the project manager alerts the necessary people and seeks a resolution that may include adjustments to the project timeline, budget, or other reallocation of resources. A project manager also works with the team to identify and mitigate risks to the project quality cost or timeline. Project management artifacts such as timelines, budgets, and operational plans are required by most research funders. Providing for project management offers confidence that research projects will be completed on time, at the appropriate quality, and on budget.

Data quality assessment is perhaps the first thing people think about when considering a data management quality system. *Data quality* is commonly conceptualized as a multidimensional concept (Wang and Strong 1996) and assessed for each dimension. The three dimensions of data quality that most heavily impact the ability of data to support research conclusions are completeness, consistency, and accuracy (Zozus et al. 2014). Weiskopf et al. (2013) described a four-part completeness assessment for research data. Accuracy can be conceptualized as a type of consistency, that is, consistency being the overarching concept that is further divided into internal and external consistencies. Internal consistency refers to the state in which data values within a data set do not contradict each other; a pregnant male would be an example of such a contradiction. External consistency refers to the state in which data do not contradict known standards or other independent sources of the comparable information; an example of an external inconsistency would be a measured temperature higher than that on record for a geographical location. Accuracy, then, is the ultimate external consistency, consistency with the true state of a thing or event of interest at the stated or implied point in time. Lesser options for external consistency comparisons are often used as surrogates for accuracy measurement when a *truthy* gold standard is not available. In practice, a researcher rarely has a source of truth and only sometimes has an independent source of the same or related information. Thus, measurement of accuracy becomes more of a hierarchy where the closeness to truth and the level of independence of the source for comparison determine the closeness of the assessment to actual accuracy (Zozus et al. 2014).

As a component of a data management quality system, data quality dimensions are defined and assessed. The assessment is often ongoing throughput a study. Data quality assessment results are then used to make corrective changes to processes when appropriate. Data quality assessment in research is rarely automated as a technical control. Most often, data quality assessment is described in procedure and operationalized as a procedural control. As described in earlier chapters, a researcher must be able to demonstrate that the data are capable of supporting the research conclusions (Chapter 1). Data quality assessments for a project are the means by which this is accomplished. Briefly, dimensions of data quality that directly impact the research conclusions are defined *a priori* (in the laboratory or project data QC procedure) along with *acceptance criteria*. Acceptance criteria are the limits for each data quality dimension, which if reached requires a specified action to be taken. Examples of such actions may include correcting data errors, fixing problems in computer programs, amending procedures, or additional training. The assessments should be systematic and objective, and the findings from the assessments should be documented. Data quality assessments are usually quantified and provided to project statisticians to assess the impact on the analysis and to jointly decide if further action is needed.

Through these assessments, the impact of data quality on the research conclusions is measured and controlled. Such assessments and associated activities, when reported with research results, demonstrate that data are capable of supporting research conclusions. There are of course situations where data quality assessments uncover problems that adversely impact the data or the study. In such cases, the benefit of the data quality assessment is identifying the problem early so that swift action can be taken, for example, additional data can be collected so that the study ultimately answers the scientific question, or in the case of fatal flaws, the study can be stopped and reinitiated with appropriate corrective actions.

The last component of the data management quality system is the management of the QMS itself. Management and maintenance of a QMS are the responsibilities of leadership. For example, if the QMS is for a lab, the head of the lab is the responsible party, if the QMS is for a research study, the principal investigator is the responsible party. Management and maintenance of a QMS include review and updating of procedures at the specified frequency as well as performing specified review and oversight functions. In addition, independent audits of the QMS can be helpful in assuring that the data management QMS is appropriately comprehensive and functioning properly. In performing such an assessment, an independent auditor will likely compare policies and procedures to applicable regulations, will assess whether the QMS covers all processes that impact data quality including those performed external to the organization or research team, will review artifacts to confirm that procedures are performed as specified, will perform a data quality assessment to confirm that the product (data) is of appropriate quality, and will trace data values from the analysis back to the source to confirm that traceability exists. Most large institutions have auditing functions. Institutional auditors can be a great resource in *formatively assessing* a new quality system for a lab or a study.

Summary

This chapter introduced the principles of quality management and provided an overview of two leading quality management frameworks. Quality management principles were distilled and applied to managing research data in a manner accessible to large and small organizations alike with an emphasis on application to labs and individual research projects. Application of these principles will help researchers anticipate, identify, and prevent, or mitigate data-related problems.

Exercises

1. A researcher has an equal opportunity (same time needed, same cost) to use a technical and a procedural control. Which is the preferable option and why?

   Correct answer: The researcher should implement the technical control. Where there is no additional cost and no adverse impact on timelines, the choice is easy because the technical control will enforce the desired procedure with the highest consistency.

2. A researcher needs to implement a procedural control. He or she has the opportunity to use a job aid in addition to the written procedure. Should he or she do this, and why, or why not?

   Correct answer: The researcher should implement the job aid. In general, job aids are knowledge *in the world*; they decrease cognitive load and associated errors by decreasing the amount of information that people need to keep in their heads.

3. On-target with minimum variance is assessed with which of the following?

   a. A measure of central tendency

   b. A measure of dispersion

   c. A measure of central tendency and a measure of dispersion

   d. None of the above

   Correct answer: (c) A measure of central tendency and a measure of dispersion. On-targetness is measured by some measure of central tendency such as a mean, median, or mode, and variation is measured by some measure of dispersion such as a range of values, variance, or standard deviation.

4. *Fitness for use, on-target with minimum variance,* and *conformance to requirements* are always different. True or false?

   Correct answer: False. They meet the same intent when customer requirements, measurement of conformance, and process management are aligned.

5. QA and QC are the same thing. True or false?

   Correct answer: False. QA and QC are different.

6. A researcher is conducting a web-based survey of professional electrical engineers registered in the state of Utah. List standard operating procedures that might be applicable.

   Correct answer: Standard operating procedures for the following would be present in a quality system for the study. (Note: these could be covered in a single document or similarly streamlined if the quality system was just for the study in question).

   Training of those collecting handling samples and data

   Documentation required for research projects

   Data collection

   Data discrepancy identification and resolution

   Data transformation and other data handling

Design, programming, and testing of computer systems used in research

Access to research samples and data

Research data quality control

Research database lock and archival

Analysis of research data

Procedures for data entry would not be required because the survey respondents are entering the data as they respond to the survey. Acceptance and integration of externally managed data would not be required, unless the survey software was managed externally.

7. What institutional policy/ies would be applicable to the study described in the question above?

   Correct answer: Correct answers should include human subjects research protection, research conduct, maintenance of research records, and research records retention.

8. An institution states in policy that the records retention period for all externally funded or published research is 5 years from the submission of the final progress report or publication of the manuscript whichever comes later. The institution has no procedures or other support for archival. What is wrong with the quality system?

   Correct answer: There is misalignment between the policy and procedural level. Because of the misalignment, record retention practices will likely have substantial variation. The policy lists no procedural control or enforcement, so the inconsistency will exist without intervention. Lastly, with no support for retention, e.g., support for storage costs, compliance with the policy relies on researchers or department allocating budgetary resources to retention activities.

9. In the above example, a researcher at the institution stated in his data management plan that research documentation would be archived in file cabinets in his or her lab. The complete set of documents is in the file cabinet 5 years later. Did the researcher comply with the policy?

   Correct answer: Yes. The policy did not state secure storage, any procedure, or any specific location. Further, the researcher stated in his or her data management plan where the documents would be stored and followed through on the statement.

10. Can a Data Management Plan (DMP) be used to document a data management QMS? If so, what should the data management plan include? If not, why not?

Correct answer: Yes. This is commonly done for projects where there is no documented QMS at the institution or lab levels. If a DMP is used in this way, it should be, at a minimum, stated as follows:

The applicable regulations and how the researcher will comply.

Procedures for all processes that manipulate or touch data

Positions and roles of staff on the research team and training provided for each

How oversight will be provided.

Computer systems used for the research, any technical controls, and software testing procedures

How the project will be managed.

How data quality will be assessed.

How, if at all, the project will be audited.

References

ASQ entry for quality assurance/quality control. Available from http://asq.org/glossary/q.html. (Accessed January 25, 2016).

ASQ entry for quality. Available from http://asq.org/glossary/q.html. (Accessed January 25, 2016).

Baker M. Irreproducible biology research costs put at $28 billion per year. *Nature News* June 9, 2015.

Baskarada S. *IQM-CMM: A Framework for Assessing Organizational Information Quality Management Capability Maturity*. Morlenbach: Strauss GMBH, 2009.

Collins FS and Tabak LA. NIH plans to enhance reproducibility. *Nature* 2014; 505:612–613.

Irizarry R, Peng R, and Leek J. Reproducible research in computational science. *Simply Statistics* 2011. Available from http://simplystatistics.org/2011/12/02/reproducible-research-in-computational-science/ (Accessed February 17, 2017).

Ioannidis JPA. Contradicted and initially stronger effects in highly cited clinical research. *JAMA* 2005; 294(2):218–228.

ISO 8000–8:2015(E). *Data quality—Part 8: Information and data quality: Concepts and measuring*, 1st ed. International Organization for Standardization (ISO), November 15, 2015.

ISO 9000:2015, Fourth Ed., Quality management systems—Fundamentals and vocabulary. International Organization for Standardization (ISO) September 15, 2015.

ISO 9001:2015—Sets out the requirements of a quality management system.

ISO 9004:2009—Focuses on how to make a quality management system more efficient and effective.

ISO 19011:2011—Sets out guidance on internal and external audits of quality management systems.

Journals unite for reproducibility. *Nature* 2014; 515.

Nosek BA. An open, large-scale, collaborative effort to estimate the reproducibility of psychological science. *Perspectives on Psychological Science* 2012; 7(6):657–660.
Peng R. Reproducible Research and Biostatistics, *Biostatistics* 2009; 10(3):05–408.
Wang RY and Strong DM. Beyond accuracy: What data quality means to data consumers. *Journal of Management Information Systems* 1996; 12(4):5–33.
Weiskopf NG, Hripcsak G, Swaminathan S, and Weng C. Defining and measuring completeness of electronic health records for secondary use. *Journal of biomedical informatics* 2013; 46:830–836.
Zhang J and Norman DA. Representations in distributed cognitive tasks. *Cognitive Science* 1994; 81(1):87–122.

14

Calculating the Time and Cost for Data Collection and Processing

Introduction

Understanding the costs associated with different data collection and management approaches is an important factor in choosing appropriate data collection and management methods. This chapter presents steps in identifying data collection and management costs, and prepares the reader to calculate resource needs and timing of those needs and to estimate a project budget.

Topics

- The hidden value in budgeting
- Modeling a research project
- The project budget
- Cost forecasting, tracking, and management

Calculating the cost of data collection and processing is a three-part process of enumerating, counting, and converting. Enumerating is the listing of human tasks and other resources needed. For common data collection and processing tasks, the task tables in Chapters 6 and 8 through 10 provide a start. Note that tasks performed by humans are listed directly and tasks performed by machines require human-performed steps of specification, programming, or other configuration and testing as well as moving the computer program from a development to test and ultimately to a production environment. Other resources such as supplies, equipment, and travel must also be listed. Services performed by others (consultants) or external organizations (vendors) must also be accounted for. Counting is the process of estimating how many of each task and other resources will be needed. For example, if handheld devices are used by data collectors, and there are two data collectors at each site, the counting entails tallying up the number of sites required for the study and multiplying by the planned number of data collectors per site. Further, the hours per data collector for the task of data collection must be estimated and multiplying by the number of data collectors. In the prior example, the conversion process entails assigning a cost per handheld device and multiplying the number of devices needed by the cost per device. Recall that the use of the devices also requires human configuration and testing steps.

The Hidden Value of Project Budgeting

The process of budgeting can have a tremendous value to the researcher. There are essentially two approaches to budgeting. One approach is to estimate a percent effort for the investigator and research team; sometimes, this can be informed by similar past projects, or sometimes, it is done as a gut estimate. This approach to budgeting does not aid the researcher much, because it does not challenge the researcher's assumptions or understanding of the actual tasks to be done in the operationalization of the research. Often, researchers do not themselves have extensive experience operationalizing research; many researchers have not processed data; recruited research participants; or specified, built, and tested computer systems. Thus, the estimating percent effort method of budgeting is limited by the investigator's knowledge and does not augment it.

The approach to budgeting advocated here is budgeting based on a model of the work to be done. Creating a model of the work to be done is a *from-the-ground-up* process that requires accounting of all of the tasks to be done and the relationships between them. The model is arrived through successively decomposing high-level categories of work to be done into more detailed homogeneous tasks. This process challenges the investigator's understanding of the operational work to be done and often requires the investigator to consult with research team members who perform the tasks in question, at the same time increasing accuracy of the model and the investigator's understanding of the work to be done. The hidden value in this process is that such a model offers an operational perspective on and can help gauge the feasibility of the proposed research. Other advantages of this approach are that (1) the level of effort is associated with the actual tasks; so if the funder does not award the full budget or if scope cuts are negotiated, the budget is easily adjusted accordingly, and the dollar adjustments correspond to the actual expected work, and (2) the model can be used throughout the project to plan and forecast the work. Over the life of the project, things change and the model provides a way to quickly see the impact of the changes on future work, effort, and costs.

Modeling Data Collection and Management for a Research Project

Consider the following study:

> A researcher is collaborating with fish and wildlife service officials on a general baseline inventory and ongoing species-specific monitoring of frogs in his state. Thirty sites across the state are included in the inventory and monitoring. To meet the general inventory as well as targeted monitoring requirements, all sites are visited during three separate weeks selected to match easily detectable times and life stages for the various species. Each site is visited four times in a visit week (four observation periods), two observation periods at night and two observation periods during the day. The visits are conducted April through August by pairs of one graduate student and one researcher. The site visit schedule works out to an average of five sites visited per month. At each visit, a data collection form is completed that includes 33 data elements and a table of frog (or other applicable life stage) observations including observation time, location, and number of individuals. Water and soil samples to be analyzed at a state laboratory are taken at the first observation period for each site. The researcher plans to have a database built to store and manage the data from the surveys and to oversee data collection and management. He or she intends to use undergraduate research assistants for data entry. One postdoc will provide day-to-day management of the study and the data processing including user testing of the database (the data postdoc) and a second postdoc will coordinate site visits (site visit postdoc). Graduate students participate in the frog study as part of a three credit hour graduate herpetology laboratory for course credit. The students commit an average of 10 hours per week in each of two consecutive spring–summer semesters. The students are made aware that some weeks will require more than 10 hours and some will be less, but the time will average to 10 hours per week and that the site visit postdoc will manage the schedule and that they will know their assignments two weeks in advance. Graduate student hours beyond the 10-hour per week course commitment are compensated at $10 per hour plus mileage through the university. The postdocs will share the responsibility of day-to-day management of the study and serving as the researcher accompanying graduate students on the visits.

Modeling the data collection and management starts with identifying the high-level tasks to be performed. In the frog study, data will be recorded on paper forms and subsequently entered into the database. The high-level tasks include (1) designing the data collection form, (2) printing field copies of the form, (3) writing and maintaining site visit data collection procedures, (4) writing and maintaining data management procedures, (5) creating and testing the database and data entry screens, (6) creating data quality checks,

(7) data collection at site visits, (8) data entry, (9) resolving identified data discrepancies, and (10) providing reports of the data. If these are not evident, one can always start with the high-level construct provided in Figure 2.3. Each task is then decomposed into tasks performed by each role. An example task decomposition is provided in Figure 14.1. For example, the data postdoc will design the data collection form using a published frog survey protocol and will draft and maintain the form completion instructions. The researcher will review and provide feedback on the design. The data postdoc will print the form on a water-resistant paper and provide water-proof pens for data recording. The site visit postdoc will draft and maintain the field procedure manual. The researcher has identified a university data management group that will build and maintain the database—they have provided their costs to the researcher. Undergraduate research assistants will perform double data entry and contact field workers (graduate students and the postdocs completing the forms) to resolve data discrepancies with review and oversight by the data postdoc. Finally, the site visit postdoc will generate status reports of the field work, and the data postdoc will generate data status reports as well as the descriptive statistics summarizing the inventory and monitoring. All individuals involved in the project will attend a weekly research meeting run by the postdocs. If data sharing is planned, after the report is generated, the data postdoc will package up the data with relevant documentation to prepare it for use by others.

The tasks above are fairly homogeneous tasks, that is, they represent doing one thing rather than a collection of different things. The goal is not to break tasks into the smallest possible tasks, but to break tasks down into the largest possible tasks that are homogenous and to which metrics can be assigned. For example, although data entry can be further decomposed all the way down to a keystroke-level model, it is a group of repetitive items that are done together toward one goal and for which it is easy to measure an estimated time per task, for example, the minutes per page for data entry rather than each keystroke. Once a list of homogeneous tasks has been created, the enumeration, that is, the base of the model is complete.

The second part of the model is the counting. For example, data entry is one task. In the above example, double data entry is used, and there are many pages to be entered. The number of pages must be counted based on the number of pages completed at each visit and the number of visits. Further, the timing of the entry makes a difference in resource availability and project planning. The counting part of the model entails counting tasks to be done per unit time; usually per month is sufficient for planning and budgeting purposes. It is more efficient to use a spreadsheet once the counting part of the model begins. An example model is provided in Figure 14.2.

Modeling data collection and acquisition are especially important when there are multiple observation points over time per observational or experimental unit, and multiple different types of data are collected.

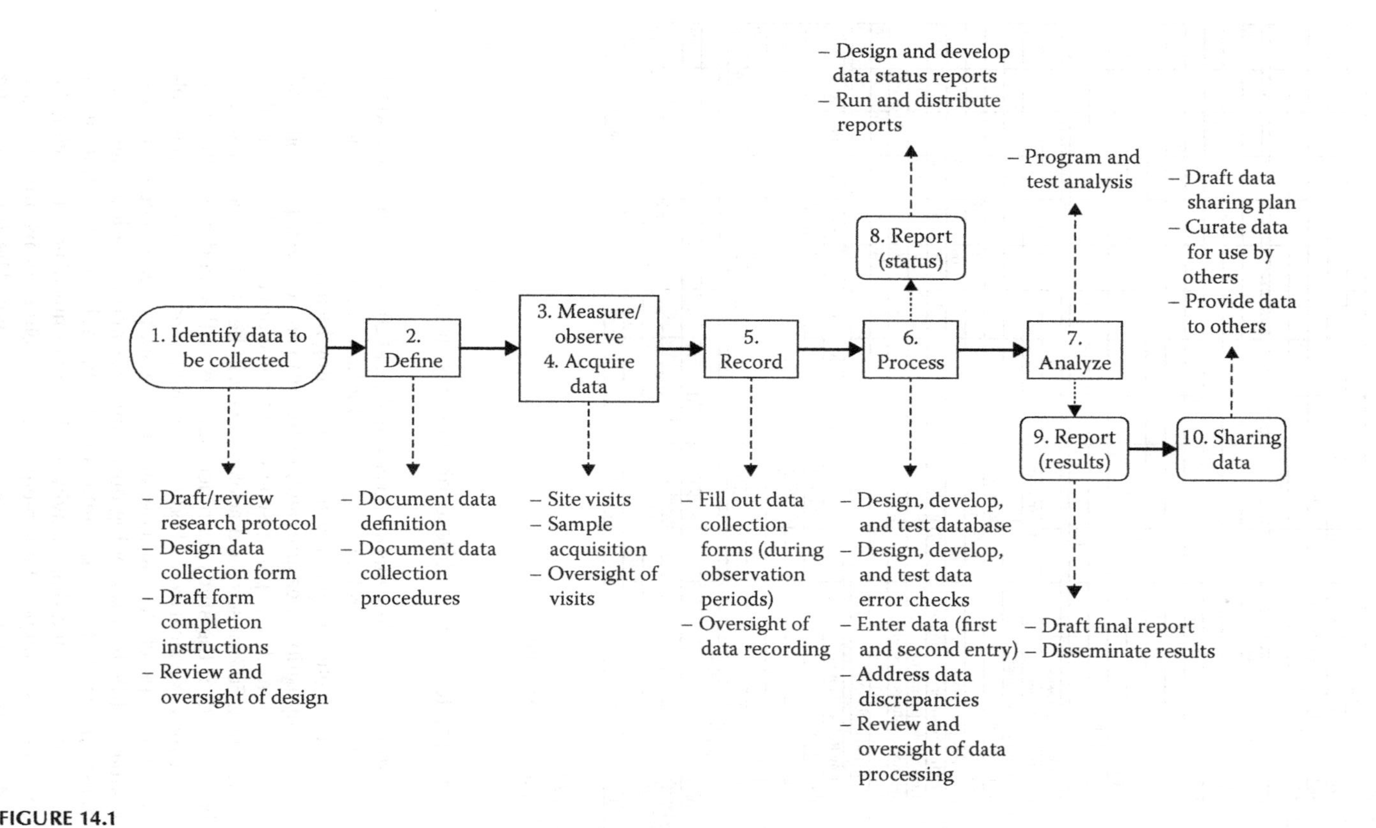

FIGURE 14.1
Example tasks associated with data collection and processing.

		Metric	Unit	Jan	Feb	Mar	Apr	May	Jun	Jul	Aug	Sep
	General Metrics											
1	Data collection form	3	pages									
2	Number of sites (field work)	30	sites									
3	Separate visit weeks per site (field work)	3	visits/week									
4	Observation periods per visit (field work)	4	periods									
5	Observation period length	4	h									
6	Data entry	0.5	h/form									
	Study Start-Up Tasks											
7	Design data collection form	5	h/page	15								
8	Draft form instructions	2	h/page	6								
9	Review draft form	1	h/page	3								
10	Print forms	3	h	3								
11	Draft site procedures	5	h	15								
12	Create site visit schedule	5	h	5								
13	Database design	10	h	10								
14	Database development	20	h	20								
15	Database user testing	15	h		15							
	Ongoing Tasks											
16	Maintain form instructions	3	h/month									
17	Weekly study leader meetings	1	h/week									
18	Maintain visit schedule	5	h/week									
19	Sites visited per month	5	sites			5	5	5	5	5	5	
20	Number of forms per month					60	60	60	60	60	60	
21	First observation of water/soil samples	2	h/site			10	10	10	10	10	10	
22	Student time on site per month		h			240	240	240	240	240	240	
23	Researcher time on site per month		h			240	240	240	240	240	240	
24	Data review	4	h/week				16	16	16	16	16	16
25	Data analysis and final report						5	5	5	10	20	30
	Key Roles—Hours			**Jan**	**Feb**	**Mar**	**Apr**	**May**	**Jun**	**Jul**	**Aug**	**Sep**
26	Principal investigator		h/month	7	4	4	4	4	4	4	4	4
27	Data postdoc		h/month	20	22	132	153	153	153	158	168	53
28	Field postdoc		h/month	30	20	145	145	145	145	145	145	20
29	Graduate student		h/month	0	0	250	250	250	250	250	250	0
30	Undergraduate student		h/month	0	0	60	60	60	60	60	60	0
	Key Roles—% Effort											
31	Principal investigator		% effort	5%	3%	3%	3%	3%	3%	3%	3%	3%
32	Data postdoc		% effort	13%	14%	85%	99%	99%	99%	102%	108%	34%
33	Field postdoc		% effort	19%	13%	94%	94%	94%	94%	94%	94%	13%
34	Graduate student		% effort	0%	0%	161%	161%	161%	161%	161%	161%	0%
35	Undergraduate student		% effort	0%	0%	39%	39%	39%	39%	39%	39%	0%

FIGURE 14.2
Modeling work to be done.

Consider the following adolescent obesity study:

> One thousand adolescents are to be enrolled at 20 clinical investigational sites. Potential research subjects attend a baseline visit where they are screened for eligibility in the study. Screening includes a physical exam, body composition measurement, exercise tolerance test, psychological assessment, and four patient-completed questionnaires. Based on pilot study data, 50% of screened adolescents will be eligible and consent to be enrolled in the study. Enrolled subjects will attend a baseline visit including a physical exam, body composition measurement, exercise tolerance test, psychological assessment, and six patient-completed questionnaires (the four screening questionnaires and two additional study questionnaires). The study intervention, behavioral counseling about eating behavior, and daily activity level occurs immediately following the baseline assessment. The screening and baseline forms are 28 pages, with an average of 25 data elements per page. The four screening

questionnaires add seven, five, five, and three pages. The two additional study questionnaires are two and three pages each. All study forms including questionnaires are entered into a web-based system. Enrolled subjects attend weekly counseling sessions where they complete the two study questionnaires (five total pages of data) and return to the clinic for quarterly medical monitoring generating 15 pages of data. Enrolled subjects use a smartphone application (optional per subject) to track daily mood, nutritional information of food consumed, daily exercise, and adverse events. Enrolled subjects use a wearable device that monitors sleep, steps, and heart rate, and use an electronic scale at home for personal optional weight measurement. The study runs for 4 years.

In the adolescent obesity study example above, the data are collected over time. Data processing and cleaning activities are planned, but can only occur when the data are available. Where study events, such as site visits or interim analyses, occur based on data acquisition milestones, such as subject enrollment or some number of subjects per site completing the first year of the study, modeling the data acquisition is useful to support scheduling in addition to tracking progress and intervening to keep the study on track. The example obesity study is further explored in the exercises.

The Project Budget

A project budget assigns costs to the resources used in the project. Budgets that are based on a realistic model of the work including both human and other resources needed allow the researcher to balance project scope and costs. Because costs are assigned to resources, a budget based on a realistic model of the work enables the researcher to adjust the project scope according to the available resources. For example, in the frog survey above, the researcher can use student enrollment in the course to project additional dollars needed for graduate student hours above the course commitment and mileage. If the overall budget is too high, the researcher has the information needed for discussion and decision making with the study sponsor about the number of sites visited and funds needed.

Budgeting for nonhuman resources, for example, equipment, supplies, registration fees, license fees, and travel, is fairly straightforward. The researcher needs only to make a list of the resources needed and an estimated cost for each. Multiyear budgeting should also take into account annual inflation, for example, 2% or 3% cost increases per year. Organizations often have budgeting guidelines for estimating cost increases over time due to things such as annual salary increases and inflation.

Budgeting for human resources takes into account additional costs of employing humans. For example, where the individuals are employees of the organization, fringe and benefits are added to the base salary (or hourly wage). Fringe and benefits can range from 20% to 30%. Where budgeting is done on spreadsheets, it is easiest to add a salary worksheet that lists the base salary for each person proposed to work on the project and the organization's fringe and benefits and annual inflation for each positions. There are often different types of positions, including temporary, students, contractors, exempt, and nonexempt employees of the institution. Exempt employees in the United States generally make a yearly salary and are exempt from the wage and hour laws, whereas nonexempt employees are paid hourly, are covered by the wage and hour laws, and receive overtime. Fringe and benefit rates often differ for different categories of employees.

Organizations also have overhead that includes costs of maintaining the organization such as office and laboratory space, utilities, central services such as human resources, information technology support, business development, and administration. Although there are norms for types of things considered to be overhead, each organization usually has methods for distinguishing overhead and project costs. Organizations in the United States that accept federal funding, for example, research grants, usually have a negotiated overhead rate, called an indirect cost rate, that is based on periodic audits. Solicitations for federal grants and contracts will specify whether the budget limit is direct costs or total costs. If the latter, the direct cost available for the project budget is the total cost minus the institution's indirect cost; thus, an institution's indirect cost rate often must be taken into account when budgeting.

Cost Forecasting, Tracking, and Management

Budgeting is an ongoing process that extends throughout the life of a project. Whether project funds are fixed or overages are allowed with approval, accurate cost tracking and forecasting are needed for successful management. Organizations that house a number of research projects usually have project-based accounting mechanisms, that is, a separate account for each project in which costs are allocated to the project. In a project-based accounting situation, it should be easy for a researcher to get up-to-date reports of costs allocated to his or her project by month. Reports that calculate over- or under-spends by month and cumulatively are often available. Managing the project budget entails matching up the work accomplished with the monies spent. It is often the case that the work occurs earlier or later than scheduled. In organizations that allocate personnel costs to projects on a percent effort basis, for example, some percent effort per month, this matching is critical. Consider the following example: Field work and associated data collection for an environmental water quality study are delayed because of difficulties identifying sites and getting permission for the field work. The researcher budgeted for approaching five sites for each one that participates. In reality, 20 sites have been approached for each of the three sites agreeing to participate. In this situation, field work and associated data collection and processing are delayed in time because of the delay in site recruitment—costs for data collection and processing would still be expected to occur as budgeted, just later in time. However, the additional time required to approach additional prospective sites (twenty rather than five for each that decides to participate) represents effort and cost in excess of what was budgeted. If the organization running the study allocated 20% effort for a project manager, 30% effort for a field site manager, and 30% effort for five data collectors, and failed to adjust allocation of effort to the project to account for the delay, the costs per month for the initial months of the study would be out of alignment with the actual work. In this case, the project management and field site management are likely at forecast or higher because of the need to approach four times as many sites as anticipated. The data collection and processing costs in this case would be very misaligned because there are no data collected yet. The appropriate action for the researcher is to reforecast the study according to the pace at which activities are actually occurring to bring the cost allocation and the actual work back into alignment as quickly as possible. Additionally, the higher number of sites approached is unlikely to decrease. Because this higher number of sites approached was not accounted for in the budget, it will become an overspend, unless cost savings can be found elsewhere in the budget or additional funds can be applied. These alternatives should be pursued as soon as the misalignment is identified and before additional work occurs. These activities are usually done by someone in the role of project manager. For small projects, the researcher himself or herself will often take on the role of project manager or assign the activities to a trainee. In large studies, an individual with training and experience in project management will usually take on these responsibilities. Budgeting for project management resources is usually expected in large studies and indicates to the study sponsor that consideration has been given to the proper management of the study. The role of the project manager is much broader than just budgeting. More information on project management can be obtained from the Project Management Institute.

Summary

This chapter introduced modeling as a tool for planning, budgeting, and managing a research project. Two examples were provided for research studies where the data collection activities were modeled and projected. Once a research study is decomposed into tasks and all of the tasks are counted, associating costs is easy.

Exercises

1. In the frog inventory and monitoring study, how many graduate students are needed?

   Correct answer: Seven students are needed to avoid providing extra compensation to the students, 250 hours per month divided by 40 hours per month; each student is expected to commit 6.25. The 10 hours per week stated graduate student commitment was estimated for a 4-week month.

2. As the investigator, what do you do if eight students register for the herpetology course?

   Correct answer: The study is the major research-teaching experience of the course, and all students should be extensively involved in the planned activities. For this reason, a good option would be to send an extra student on some site visits and use the additional data as independent observations over the same period, that is, additional quality control.

3. What is the impact to the study if only four students register for the herpetology course?

   Correct answer: Four students leave 90 hours per month needed (250–4 × 40 h/month). The 90 hours per month of work will have to be paid at the agreed rate of $10 per hour. Thus, the cost to the study budget is $900 per month plus mileage to compensate the four registered students for the extra site hours.

4. Row 19 of the frog study model shows five sites visited per month. How is the 250 graduate student hours per month calculated?

   Correct answer: Five sites visited/month × three separate visits/site × four observation periods per visit × 4 hours per observation period + 10 hours for first visit soil/water sample collection.

5. What happens if the weather, seasonal queues the frogs follow, and student availability differ from five sites visited per month?

   Correct answer: The only thing the research team can be sure of is that the actual timing of the visits will differ from the estimated five sites visited per month. This is why one of the postdocs is assigned to manage the visit schedule and assignment of students to visits. Resource models and budgets are estimates and can only go so far. A project manager is the connection between the plan and the reality. The role of the project manager, in the case of the site visits, the fieldwork postdoc, is to track the visit schedule and actual visits and manage the schedule, for example, if seven sites have to be visited in 1 month to work that out with the graduate students.

And to assure that all visits happen whether or not they cross a month boundary.

6. Another investigator approaches the frog study investigator. The new investigator is interested in inventorying salamander populations in the frog study observation sites. The salamander inventory form is three pages long and has different data than the frog form, and the observation periods are at different times of the season. How might the salamander inventory be budgeted?

   Correct answer: The fact that the site visits are timed differently means that the same site visits cannot be used for both studies. Similarly, the salamander data collection form contains different data, so the data collection procedures and database are different. Thus, aside from the soil and water samples which can be used for both studies, the salamander study might use the same budget template but would cost almost as much as running the frog study.

7. The frog study sponsor approaches the investigator with an idea. The sponsor would like to broaden the frog inventory to include all states in the southeast region in the United States. What high-level information is needed to estimate the cost?

   Correct answer: The species monitored and their cycles in each area determine when visits need to occur. Depending on the data from prior years of the Frog inventory in one state, they may decide that fewer visits per site would be acceptable, and that fewer sites per state may be acceptable to gain a larger geographical area of the survey. The timing of the site visits, the number of sites visited, and the number of visits per site are all major cost drivers. In addition, travel to the locations in other states may increase cost, unless additional collaborating institutions can be brought into cover sites closer to them.

8. An investigator for a study is estimating the cost of a proposed project. The investigator presents the budget as percent effort for the roles involved. When asked how he or she arrived at the level of effort, he or she explained that he or she did it based on how much of a person she thinks is needed. What is wrong with this approach?

   Correct answer: The estimate given was not dependent on the actual work. If/when asked to reduce the budget, the investigator would have no way of matching reduced budget to the corresponding ways that the scope would need to be reduced.

9. Under what conditions would the approach described in question 8 be acceptable?

   Correct answer: If the investigator had conducted several studies that were very similar in operational design, and the estimates were based on actual expended effort in those prior studies, the estimates would likely be accurate.

Create a spreadsheet model of enrollment and data collection for the adolescent obesity study, and use your results to answer questions 10–20.

10. Based on your model of the adolescent obesity study, if the sites are ready to enroll subjects in the following way (month 1—one site, month 2—two sites, month 3—four sites, months 4 and 5—five sites each, and month 6—three sites), and each site screens 20 patients per month, how many patients are screened in month 7 across all sites?

    Correct answer: 400 patients are screened (see the adolescent obesity study model).

11. Given 20 sites screening 20 patients per month and an enrollment rate of 50% (half of the patients screened actually enroll), how many patients are enrolled in month 3?

    Correct answer: 70 patients are enrolled (see the adolescent obesity study model).

12. In what month does the study meet its enrollment goal?

    Correct answer: Month 8 (see the adolescent obesity study model).

13. Assume that the study enrolls an additional 200 subjects over the enrollment goal of 1000 subjects and screening data come in one month after screening, and baseline data come in the following month. How many pages of data from baseline forms are collected in month 8?

    Correct answer: 11,600 pages of baseline data (see the adolescent obesity study model).

14. In what month does the last baseline form come in?

    Correct answer: Month 11 (see the adolescent obesity study model).

15. How many total pages of screening data are expected for the study if additional 200 patients over goal are enrolled?

    Correct answer: 115,200 pages (see the adolescent obesity study model).

16. How many total pages are collected for the quarterly medical monitoring visits if additional 200 patients over goal are enrolled?

    Correct answer: 270,000 pages (see the adolescent obesity study model).

17. How many total pages of data are collected for the study if additional 200 patients over goal are enrolled?

    Correct answer: 1,606,800 (see the adolescent obesity study model).

18. If data entry takes 2 minutes per page and a professional data entry operator hired by each site makes $15 per hour, how much can this study expect to spend on data entry alone?

    Correct answer: (1,606,800 × 2 min/page) = 53,560 hours. At $15 per hour, data entry will cost the study $803,400.

19. If additional 200 patients over goal are enrolled, how many data transmissions from devices should the study expect per month at full enrollment?

    Correct answer: 76,800.

20. Given the large number of data transmissions, suggest a way to identify potential problems, for example, a device that is not sending data, or aberrant data, for example, a patient's dog on the scale transmitting data, or an accelerometer left in a car.

    Correct answer: There are many variations of economical ways to do this, and different amounts of checking; so there is not one correct answer. However, all answers should acknowledge that human inspection of all transmissions is not feasible. An example of a monitoring scheme may be to define 50 or so rules that check for things like the examples above, for example, changes in a data stream from a subject's device which are not likely due to natural variations. These rules run on transmissions as they are received. Any rule that fires generates an electronic flag to the clinical investigational site. The site checks with the patient and works to resolve the issue. A more efficient way may be that the rule first generates a message to the patient's smartphone, and if the data stream does not come back into range, a flag to the site is generated and so on.

21. How could the data collection model for the adolescent obesity study be used to manage the study?

    Correct answer: The projected number of pages and data transmissions per month is the estimate of what should be received. Expected versus actual can be checked monthly or weekly and divergence of the two should be investigated and used to improve the data collection and submission processes. Further, reports of late data, for example, 2 weeks late or 30 days late, can be generated by subject and by site. The late reports can be used to work with the sites to improve, and shared with all sites, and study leadership to promote keeping data collection current.

15

Research Data Security

Although the type and level of security necessary vary with project and institutional needs, every project needs some level of protection against data damage and loss. Where privacy and confidentiality rules apply or sensitive data are concerned, higher levels of security are necessary. This chapter covers security considerations for protection against data damage and loss as well as special considerations for sensitive information. Concepts relevant to securing research data are presented to guide researchers in identifying data security measures needed for specific research situations.

Topics

- Security as protection against loss of confidentiality, integrity, or availability
- Steps investigators can take to prevent loss
- Risk-based categorizations of data and information systems
- Implications for contractual data protections
- Components of a research data security plan

Security as Protection against Loss of Confidentiality, Integrity, or Availability

Securing research data protects humans and organizations participating in research (if any), research funders, the investigator, and the investigator's institution. Often the phrase data security connotes some robust information technology (IT) infrastructure with cold rooms, raised floors, and lots of wires maintained by skilled professionals watching over racks of expensive computers with specialized software in locked rooms. Although in many cases, this scenario is necessary, can be leveraged by, and benefits researchers, it is helpful to think about data security in a simpler, broader, and research centric way. Securing research data in some ways is not much different than protecting personal information against identity theft and insuring and securing a home. From the standpoint of an investigator, research data security is about reasonable protection against damage and loss. The Federal Information Security Management Act of 2002 (FISMA) defines three types of loss: loss of confidentiality, loss of integrity, or loss of availability (44 U.S.C., Sec. 3542). A *loss of confidentiality* is defined as the unauthorized disclosure of information. A *loss of integrity* is defined as the unauthorized modification or destruction of information. A *loss of availability* is defined as the disruption of access to or use of information or an information system (FIPS Pub 199). The approach here takes a similarly broad perspective on the data to be secured and considers the data itself, metadata and other documentation necessary to understand and use the data, and operational artifacts documenting specific details of data management such as the audit trail. Data security is composed of the managerial, technical, and procedural controls put in place to protect against loss of confidentiality, integrity, or availability of data. The following scenarios illustrate the types of loss and the resulting damage:

Take the common scenario of a researcher and a graduate student publishing a paper based on work that the graduate student did in the researcher's laboratory. The graduate student collected and analyzed the data or her. Two years after the graduate student leaves, the researcher receives an inquiry about the paper. A reader is wondering whether the labels on two lines in a graph could have been reversed, that is, the lines in the figure may be mislabeled. In order to answer the reader's question, the researcher needs to look at the data from which two figures were produced. However, the researcher does not have the data and remembers that the student processed all of the data on his or her personal laptop. The researcher is able to contact the student, but the student reports that the laptop had a catastrophic failure and has been recycled. In this hypothetical but common scenario, the researcher ends up explaining this to the reader and proactively to the journal. Situations like this where findings are drawn into question and the question cannot be resolved often end in an expression of concern or retraction of the paper.

Another researcher similarly works with a small team to complete a study. Two team members collected and analyzed the data. Four years later, the researcher is doing related work and wants to combine the old data with some similar and recently collected data. When the researcher looks for the data, the old data are not on the laboratory computer system where they should be. The researcher is able to contact the students. Neither of them have a copy of the data, and both report using the appropriate shared directory on the laboratory computer system for all of the data processing and analysis work. All of the students who work or have worked in the laboratory have or have had access to the system. University IT department backs the system up every 30 days, but the backups are only kept for 6 months, and the needed files are not on the available backups. Without the data, the researcher could not do the pooled analysis. These two scenarios, while made up, are illustrative of common data security problems and lack of rudimentary procedures to assure that investigators have access to the data over a reasonable period of time. Institutions with policies stating that data are owned by the university and stewarded by the investigator imply an expectation that investigators have and monitor policies for maintaining and documenting data during a project and archiving data with associated documentation afterward. However, most investigators do not have procedures in place to dictate whether and if so, how data are organized, documented, and archived during and after a project.

The above security-related scenarios require no special knowledge on the part of the investigator other than to set up a process using commonly available organizational resources, to communicate those processes to those working in the laboratory, and to monitor compliance with them. In these cases, no special security knowledge is needed. The following examples, however, highlight the investigator's responsibility for data security even when more specialized and advanced methods and technology are required for securing data.

In the summer of 2009, the Carolina Mammography Registry (CMR) was *hacked*, and the records of 180,000 study participants were exposed, including 114,000 social security numbers (UNC 2011). The server was owned by the University but operated by the CMR study. After the breach was discovered, University administrators initially decided that the investigator, a professor of radiology at the medical school with an admirable academic record, who had been running the study for over 10 years without incident, had "been careless with sensitive information and had damaged public trust in the university, and should be terminated" (Kolowich 2011). A faculty hearings committee later persuaded the university instead to demote her to associate professor, allow her keep her tenure, but cut her pay by half. The investigator appealed the disciplinary action, and the matter was settled almost 2 years later when the investigator and the University agreed that the appeal would be dropped; the investigator would be reinstated with the salary cut rescinded but would retire from the University later that year (UNC 2011). Both parties agreed that the investigator "had the responsibility for the scientific, fiscal and ethical conduct of the project, and responsibility to hire and supervise the CMR information technology staff" (UNC 2011). This case sparked a significant debate about an investigator's responsibility for data security, especially because most investigators are not trained or skilled in IT and often are not in a direct supervisory role over such staff.

A few years earlier, a laptop with social security numbers was stolen from the locked trunk of a National Institutes of Health (NIH) employee (Weiss and Nakashima 2008). The government computer contained personal and medical information of 2500 patients enrolled in a NIH study. According to the government's data security policy, the laptop was in violation, because it should have been but was not encrypted. The press release reported that an initial attempt to encrypt the computer failed, and the employee, a Lab Chief, did not follow up; thus, at the time of the theft, the data stored on the machine were not encrypted (Weiss and Nakashima 2008).

Steps Investigators Can Take to Prevent Loss

Any of these situations such as data not stored in an appropriate area, inadvertent deletions, changes to data, a computer failure, attempted internet intrusions, or theft of a mobile device can happen to any investigator. However, there are systematic steps that investigators can take to protect research participants, research data, and research reputations. The first is to use data systems that automatically track changes to all data values in a way that the tracking cannot be subverted by a system user, for example, by using technical controls built into the data system to track changes to data. Such tracking can be periodically monitored for unusual behavior and addition of double data entry, double coding, or double transcription as described in Chapters 6 and 8 guards against sloppiness. The second precaution is to compare a sample of data from analysis data sets back to the record of the original data capture or to compare counts between the two. It is easy for data to be inadvertently excluded, the wrong or test data to be included, or data values to be *hard coded* in computer programs written to do the analysis then forgotten. Audits and counts help detect such trouble as does independent programming of important analyses. All of these add an element of independent review and, in doing so, strengthen the processes by which data are handled. The resulting data are less vulnerable to mistakes or untoward changes.

The next group of precautions was discussed thoroughly in Chapter 13 on data management quality systems and includes managerial controls. These are very applicable to data security, for example, procedures for managerial approval for access granting when new team members join or existing team members take on new assignments and automatic access revoking when members of the research team leave the study. Additional managerial controls relevant to data security include review of system logs, review of directories or systems housing data, computer programs to confirm compliance with file management or data storage procedures, periodic confirmation or testing of backups, and existence of position descriptions, training with management oversight of work performance. For oversight to be effective, it must be provided by someone with equal or more experience in an area than the supervised individual. This does not have to be the investigator, but it is the investigator's responsibility to assure that individuals are qualified for the tasks they perform and that appropriate oversight is provided.

Similarly, the investigator is responsible for identifying the level of security protection needed for the data and for providing that security. This does not require an investigator to be a data security expert but does require an understanding of applicable institutional policies and consultation with institutional experts for assistance in planning for appropriate data security. This planning is best done at the research proposal and funding stage, so that the project is appropriately resourced for the needed protections.

Although detailed information security methods and technology are very much beyond what can be provided here, the following framework provides a way for investigators to think about some of the security approaches that are required for different data scenarios. Consider the different combinations of data moving (or not) during a project and different types of devices on which data may be housed. Data can move over the Internet or other communication links, or data can be on physical media that move. Data can be stored on any number of mobile devices or on computer systems within institutions. Such computer systems can be open to (accessible from) the Internet or not. There are two basic variables; whether data move and whether the computer system in which the data are housed is open to the internet (Figure 15.1).

The case of data at rest and a system that cannot be accessed by the Internet (top left quadrant in Figure 15.1) is the simplest. Examples of data storage strategies in this quadrant include paper forms, and desktop computers and servers not accessible from the Internet. If the investigator controls access to the system and the space in which the system is housed, the data within the system are secure. Protections may be needed to prevent data from being written out of the system and e-mailed, saved elsewhere on an internal network, or saved to media that can be removed. Where the system is confirmed to be in good working order and maintained as specified by the manufacturer, system access is controlled, actions taken within the system are attributable to individuals (this is another way of saying that there is an audit trail), the audit trail is monitored for oddities, and backups are made, checked, and stored; the data stored on the system are relatively safe. There must also be a clear plan for data documentation, preservation, and archival after the project. Although all of these actions can be taken by investigators

	At rest	Moving
Closed system	Physical security	Physical security for stand-alone mobile devices, securing device and data against physical intrusion
Open system	Physical security, securing data against instrusion	Physical security for stand-alone mobile devices, securing device and data against physical and Internet intrusion

FIGURE 15.1
Researcher's framework for indicating data security needs.

and research teams. Security requirements for most federally regulated work and for some grants and contracts can be complex, and it is best for the investigator to have security measures reviewed by an institutional information security officer to assure that the measures are appropriate for the project and meet any regulatory requirements, funder requirements, data use agreements, or institutional policies.

The bottom left quadrant, data are also at rest, but in a system that is open to the Internet. This situation requires the precautions above as well as additional measures to protect against unwanted access from the Internet. These might include hackers trying to access data or perturb the system operation in some way. Internet-based attacks come in many forms; securing against them is complex and should be handled by a qualified IT professional specialized in securing systems open to the Internet (this includes on a network that is open to the internet). In addition, it is best for the investigator to have security measures reviewed by an institutional information security officer to assure that the measures are appropriate for the project and meet any regulatory requirements, funder requirements, data use agreements, or institutional policies.

The top right quadrant is a system that is not accessible from the Internet, but where data are moving. Examples that fall within this quadrant include data mailed or otherwise transported on digital media, for example, mailing a hard drive, a computer, or mobile devices used in the field to collect data. Data on such mobile media require the precautions described above for the quadrant to the left, as well as additional security measures against physical intrusion during transport, and measures to protect stored data in the event that the device falls into the wrong hands. Such additional measures include controlling access to the device, destruction of data after a number of failed logins, and encryption of data stored on the device. There are many different approaches to securing data on media and devices, including selection of more secure media and devices. The approaches and device functionality often vary in nuanced ways. Selection of security approaches is best handled by a qualified IT professional specialized in securing data on media and devices. In addition, it is best for the investigator to have security measures reviewed by an institutional information security officer to assure that the measures are appropriate for the project and meet any regulatory requirements, funder requirements, data use agreements, or institutional policies.

The bottom right quadrant shows data stored on a system open to the Internet where some or all of the data are moving via the Internet, other communication links, via portable media or mobile devices. The framework does not distinguish between data being transferred over the Internet and data being carried on a device (some portable media or a mobile device), because data being sent over the Internet are stored on some device somewhere, usually more devices and more somewheres than we would like. The scenario in the bottom right quadrant is common today and requires all of the aforementioned protections. No matter which quadrant a particular project or parts of a particular project inhabit, the investigators should always have access to and control of the data. Further, different parts of a project may fall into different quadrants. For example, a study in which some data are collected on paper forms and other data are collected in the field and sent via the Internet from the devices in the field would occupy at least two quadrants. In addition, a project may inhabit different quadrants during the data life cycle, for example, data may be collected on smartphones configured for a study that are not open to the Internet; essentially, data are managed on a closed system. Then after the study has ended, data are shared publically over the Internet where the investigator may wish to assure that they are not adultered.

Often security is juxtaposed with accessibility, for example, the perception that more secure systems are less accessible, and systems that are accessible are not secure. The ideal is that systems at the same time will be easily accessible to those who should have access and impossible to access by those who have not. Today, in most areas, this ideal remains out of reach. Most of the time, our technology, methods, and implementations blunt the desired sensitivity and specificity of security approaches, for example, forgotten passwords prevent system access when it is needed by the user, or the effort associated with logging in to a system presents a barrier to use. These issues are unintended side effects of securing systems; they can and will improve over time with methodological advances and collaboration between researchers and IT professionals.

In addition to that part of security which becomes a barrier to system use, the costs of security must also be considered. Managerial and procedural controls require people to allocate time to security-related tasks such as reviewing system logs and reports for oddities, providing oversight to employees, confirming system access requests, and confirming conformance to file naming and storage procedures. IT professionals required to implement and maintain software and hardware aspects of security also require allocation of their time and funds to cover the software and hardware costs. For this reason, a risk-based approach to data and system security is often taken by institutions.

Risk-Based Categorizations of Data and Information Systems

Most large organizations have policies, procedures, and infrastructure to classify data to be secured. Such classification systems encode an institution's identification and assessment of risks associated with data and dictate the level of security expected based on where a project's data fit into the classification. Often these classification systems take into account many different types of risk to an institution, including regulatory compliance, legalities associated with privacy, confidentiality of individuals, patent law, institutional reputation, institutional intellectual property procedures, and contractually bound confidentiality. There are many such classification systems. For example, the National Institute of Standards and Technology (NIST) Guide for Developing Security Plans for Federal Information Systems uses a three-level classification, categorizing information systems storing or processing data according to the FIPS 199 categories (NIST/US DOC 800-18 2006, FIPS 199 2004). The FIPS 199 categories form a three-by-three matrix of the three types of security breaches (types of loss) and three levels of potential impact on organizational operations, organizational assets, or individuals should such a breach occur (FIPS 199 2004). Academic institutions seem to favor one-dimensional four- or five-level classification systems based on impact and regulatory requirements. Figure 15.2 shows a five-level classification system that was compiled by generalizing across similar classification systems from several academic research institutions.

Reading from left to right, Figure 15.2 classifies data as low risk (public information), mild risk, moderate risk, high risk, and very high risk. Note that there is not a category indicating no risk. Even data which are made publicly available carry a level of risk. Such risks may include malicious adulteration of data on an organization's Website or data that contain simple but embarrassing errors such as incorrect citations, outdated information, broken links, or misspellings. The low-risk category (Figure 15.2) includes information that are shared publicly outside the institution. Precautions are usually taken to protect these data from unintended adulteration and from malicious activities attempting to render them inaccessible. The second category, mild risk, includes data that are not legally protected, that is, not protected by laws, regulations, or contractual arrangements but which may be considered

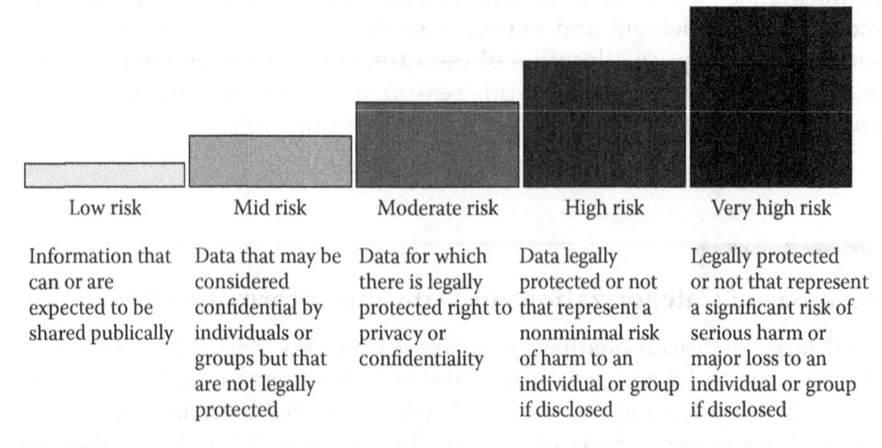
Low risk
Mid risk
Moderate risk
High risk
Very high risk
Information that can or are expected to be shared publically
Data that may be considered confidential by individuals or groups but that are not legally protected
Data for which there is legally protected right to privacy or confidentiality
Data legally protected or not that represent a nonminimal risk of harm to an individual or group if disclosed
Legally protected or not that represent a significant risk of serious harm or major loss to an individual or group if disclosed

FIGURE 15.2
Example for risk-based classification of data.

confidential by individuals or groups. For example, documents supporting research in progress are often considered confidential by investigators and research teams who perceive risk to the study conduct, for example, if the blind were broken, or if details of the study design became available to competitive researchers who might use the information to advantage to further their own work. In 2008, a web portal used by investigators and researchers working on a large multicenter clinical trial was accidentally exposed when the password protection for the third-party hosted trial web portal lapsed. An individual external to the trial team publicly accessed, archived, and publicly shared the entire portal and its contents making the study design documents, trial manuals, procedures, and contact lists publicly available. Although no information about trial participants was stored in the web portal, the unintended disclosure of the study materials cost those involved with the study time (Barrett 2012).

The third category (Figure 15.2—moderate risk) is designated as those data for which there is a legally protected right to privacy or confidentiality. Data in this category may include human subjects participating in clinical research involving protected health information or data that are contractually held confidential by data use agreements between parties. These data represent additional risks for organizations because disclosure also means lack of compliance with some legal requirement. The fourth category, high risk, includes data that represent a nonminimal risk of harm to an individual or group if disclosed. Examples include social security numbers, financial account information, and medical information not otherwise classified as very highly risky. These data may or may not be legally protected. What makes them highly risky is the potential for harm to an individual or group. The very high-risk category includes data that represent a significant risk of serious harm or major loss to an individual or group if disclosed. Examples of very high-risk data include information that carries a significant risk of criminal liability, could cause significant psychological harm or other injury, and would jeopardize insurability or employability. Similar to the high-risk category, these data may or may not be legally protected. Note that different levels may apply to different types of data collected on a project and that some of these may also apply to research records or data provenance information not just final data.

Not all institutions appraise risk in the same way, for example, some include civil or criminal liability in the higher risk categories. Some categorization systems are weighted more toward harm to individuals or groups such as the example in Figure 15.2, whereas others are weighted more heavily toward regulation and legal aspects, while others are weighted toward risk to the institution whether from harm to others, from reputation damage, or loss of market share, and so on. Because these classification systems support institutional requirements and procedures, investigators should maintain familiarity with them and with institutional resources that can be consulted for guidance or relied upon for support to assure compliance.

Such institutional categorizations, whether they apply to just research data or more broadly across an institution, usually take into account multiple stakeholders. For example, institutional compliance with governmental regulations, such as the Health Insurance Portability and Accountability Act, the Family Educational Rights and Privacy Act (sometimes called the Buckley Amendment), and the FISMA. Safeguards required by research funders can also require compliance with any of the above or custom requirements. Further, institutions often have their own requirements to safeguard information assets and mitigate risk.

Implications for Contractual Data Protections

An investigator may or may not own research data; data ownership is further discussed in Chapter 16. Where a researcher does not himself or herself own the data, he or she may not have the authority share or otherwise further disclose data or enter into such an agreement. In either case, if research data are the subject or a subpoena, national security request, or court order demanding disclosure of information in their possession, the researcher should consult (1) his or her institution's legal counsel regarding his or her authority to respond to the disclosure request and the procedures for doing so, and (2) where human subjects are involved, the institutional ethics committee.

For certain kinds of data, generally research data containing individually identifiable information, researchers can apply for government protections to guard against forced disclosure. These protections are available for subjects if the research data could damage their financial standing, employability, insurability, or reputation. For example, a Certificate of Confidentiality in the United States will allow a researcher to refuse to disclose personally identifiable data concerning research subjects in civil, criminal, administrative, legislative, or other proceedings, and can be obtained from the NIH for NIH-funded research. The Department of Justice may grant similar protections for research involving criminal records, and research funded by the federal Agency for Health Care Policy and Research is offered protection.

Data use agreements or other contractual arrangements for using and sharing data usually require a transitive enforcement of the protections for further disclosure or explicitly prohibit it. For example, if institutions A and B enter into a data sharing agreement by which institution B can use a data set provided by institution A, such an agreement would either explicitly prohibit institution B from any disclosure of the data or would require the institution B to hold anyone using the data, for example, independent contractors working with institution B, to the same protections as those to which institution B is held by the data sharing agreement with institution A. Thus, anytime contractually protected data are accessed by those other than the parties to the original agreement; an additional agreement may be required if such use is permitted at all. In the case of contractually shared data, the data are often subjected to the policies of both the sharer and the sharee's institution.

Components of a Research Data Security Plan

Some institutions require researchers to document a security plan for research data. Some funders require research data security plans either as stand-alone documents or as part of an overall data management plan (Williams et al. 2017). Such a plan should identify the data elements to be processed or stored, the type(s) of media on which they will be stored, and the names of all software used to store, process, or access data. For each software or information system used to process or store data, whether the system is accessible from the Internet, all planned data transfers into or out of the system, and a maintained list of everyone with access to the data and the type of access (read, create, delete, update), a description of how the software was validated, change control procedures, logging and log monitoring procedures, how users and interfacing systems are authenticated, agreements and descriptions for direct system interfaces, a description of the security and backup procedures for data in the system, and a statement regarding data encryption. In addition, elements of physical security that apply to specific systems or to the project as a whole should be described. Managerial controls applied to specific information systems or to the project or institution in general should be referenced such as *rules of behavior*, procedures for *separation of duties*, onboarding and termination procedures, and procedures for performance monitoring and managerial oversight. For an organization of a large research laboratory with a stable information system, it may be most efficient to document many of these aspects in laboratory-wide documentation rather than for each project. If the research conducted requires a stand-alone data security plan or security information to be specified in the data management plan, institutional information that does not change often can be included in plan templates to decrease the documentation burden for each project (Table 15.1).

TABLE 15.1

Example Content of a Research Data Security Plan

A list of all data elements and their data sources
An indication or classification of level of risk of each type of data
Places including media types where data will be stored
Name and role of the individual managing the infrastructure where electronic data will be stored
Software used to collect, manage, and store data
Planned and foreseeable data transfers into or out of the system
List of and agreements regarding direct system interfaces
A list of those able to access data and their assigned privileges (create, read, write, and delete)
A description of physical security relevant to the system
Technical, procedural, and managerial controls relevant to system security. Such controls may include the following:
Technical controls
Encryption and sanitization
Communications and host-based security
Logging and system monitoring
Backups and disaster recovery
Procedural controls
SOPs describing system development, maintenance, and use
User training
Managerial controls
Rules of behavior
Procedures for separation of duties
Onboarding and termination procedures, for example, access provision and revocation
Procedures for performance monitoring and managerial oversight
Software validation and change control

Summary

This chapter approached data security very broadly, from the perspective of protection against damage and loss. Damage and loss can either occur to data directly or be experienced by human research subjects, investigators, research teams, and institutions as a result of damage or loss to data. Generalized and specific examples of damage and loss were provided. Systematic actions that investigators can take to prevent damage and loss were discussed. A framework was provided to aid in determination of types of security that might be needed. An example risk-based classification scheme was explained. Implications for contractually protected data and sharing or use of data owned by others were discussed. The chapter is closed with a description of common items considered in writing a research data security plan.

Exercises

1. Which of the following is/are among the types of loss described by federal information processing standards?
   a. Privacy
   b. Availability
   c. Intrusion
   d. Denial of service

   Correct answer: (b) Availability. The other two types of loss described by FIPS Pub 199 are lost of confidentiality and loss of integrity.
2. A denial-of-service attack is best characterized as which type of loss?
   a. Privacy
   b. Integrity
   c. Availability
   d. None of the above

   Correct answer: (c) Availability. The purpose of a denial-of-service attack is to decrease availability of the data or other resources of a computer system.
3. A study is collecting data on paper forms after which they are entered into the study database. Other data in the field are captured by study participants using a smartphone application that they download for free from the study's Website; data captured on the device are sent via cellular service to the main study database without storage on the device. Which quadrant(s) of the framework best characterize the study?
   a. Top left only
   b. Top right only
   c. Top and bottom left
   d. Top left and bottom right

   Correct answer: (d) Top left and bottom right. Paper forms fall into the top left quadrant. Use of the mobile device is irrelevant because no data are stored there. The data from the mobile device, however, do move over a cellular telecommunication link; although it is not the Internet, a cellular link indicates an open system.
4. Data for a thermal monitoring study are collected in the field using thermal cameras. The cameras are connected to computers where the images and their date/time stamps are stored. Once a week, a research assistant travels to each site, tests the equipment,

and swaps out the portable drive with a new one. The drive with the field data is taken back to the laboratory for processing on an internal computer system. Which quadrants of the framework best characterize the study?

a. Top right
b. Bottom right
c. Bottom left
d. None of the above

Correct answer: (a) Top right. The machines in the field and the portable hard drives are mobile devices that store data. There is no indication that the system is open to the Internet or other telecommunication links or that data move over any such links.

5. A researcher is running a study and four members of the laboratory need to enter data into the study database. Which of the following is the best option?
   a. Have each person enter data into a separate copy of the database on his or her computer.
   b. Use a data system controlled by the researcher with user authentication and an audit trail
   c. Have one person enter all of the data in the researcher's computer.
   d. It does not matter how each person enters the data as long as the researcher has a final copy

   Correct answer: (b) Use a data system controlled by the researcher with user authentication and an audit trail. Option (a) leaves the investigator at risk because data are on individually controlled computers rather than a system controlled by the investigator. Option (c) is a possible solution to enforce accountability of changes to data but incorporates no tracking. Option (d), like (c), lacks tracking of entries and changes made to the data during data processing. Thus, option (b) is the best because data entries and changes are tracked to each person and the entry is done on a system presumable in the investigator's control.

6. At an institution that owns research data, a postdoc completes his or her work and is transitioning into a faculty position at another institution. The postdoc would like to have a copy of the data for the research that he or she and his or her mentor did together. Which of the following is the best option?
   a. The postdoc cannot have a copy of the data because they are owned by the institution.
   b. The postdoc and mentor decide to address data access after her transition.

c. The postdoc and mentor draft and sign a data use agreement stating that the postdoc can maintain continue to use a copy of the data

d. The postdoc and mentor draft a data use agreement stating that the postdoc can maintain and continue to use a copy of the data and take the draft agreement to university legal counsel for an authorized official to review and sign

Correct answer: (d) The postdoc and mentor draft a data use agreement stating that the postdoc can maintain and continue to use a copy of the data and take the draft agreement to university legal counsel for an authorized official to review and sign. Although (a) may be the result of an attempt at (d), resorting to (a) without trying is not fair to the postdoc. Option (b) leaves the postdoc at risk of not being able to access the data later. Option (c) at an institution where the institution owns the data means that the researcher is likely breaking the rules and signing a document that he or she is not authorized to sign.

7. According to the five-category scale, which of the following would be low-risk data?

   a. Information about the health conditions of research participant's family members

   b. Information about research participant's use of street drugs

   c. Research participant's anonymous opinions collected in a food taste test

   d. None of the above because all data are collected from human subjects

   Correct answer: (c) Research participant's opinions collected in a food taste test. The taste test responses pose only minimal risk of harm to research participants and to the organization if disclosed and are not protected by law or regulation.

8. Publicly shared data on an institution's Website require no security. True or false?

   Correct answer: False. Publicly shared data are vulnerable to loss of integrity through malicious alteration and loss of availability through denial-of-service attacks.

9. A private foundation contracts a university laboratory to conduct a study and requires compliance with the components of FISMA in the contract. The contract is fully executed by the University signing official. Which of the following is true?

   a. The University is not bound to FISMA because the foundation is not a federal agency.

   b. The University is bound to FISMA because it was agreed in the contract.

c. The researcher is not required to comply with FISMA because she did not sign the contract.
d. FISMA does not apply because the University is not a federal agency.

Correct answer: (b) The University is bound to FISMA because it was agreed in the contract. Contracts between parties can rely on external standards such as federal regulations.

10. A data system for a research study is breached, and several hundred records are disclosed. Because of the type of data, a federal fine is imposed. Who is responsible?
    a. The University if the study is funded through a grant to the university
    b. The University signing official who signed for the grant
    c. The IT professionals and management overseeing the server
    d. The researcher

    Correct answer: (d) The researcher, because most often University policies hold the principal investigator accountable for a study's conduct.

11. A research data security plan must be a stand-alone document. True or false?

    Correct answer: False. A research data security plan may exist as a stand-alone document, but in most cases, there is not a requirement for it to be so.

12. A researcher undertakes a project and submits a grant for funding. In the budget, the researcher estimates the IT-related costs for information security; the project involves high-risk data, so he or she knows that some security will be required. Afterward, the researcher approaches IT to provide the needed security and discovers that the cost is well over what he or she initially budgeted. Which of the following are possible courses of action?
    a. The investigator accepts the risk and proceeds without the needed security.
    b. The investigator works with IT to alter the data collection and management strategy constraining the high-risk data to paper forms completed and stored in a secure room on campus.
    c. The investigator admits his or her mistake, approaches his or her department chair for budgetary help to resolve the situation if such help is not available, she is prepared to decline the grant.
    d. All of the above.

    Correct answer: (d) All of the above are possible courses of action. (b) and (c) are the better options. (b) is not always possible.

13. What other options might be available to the researcher in the question above to resolve the situation?

    Correct answer: There are probably lots of options. Some include fining another investigator collecting similar data under appropriate security controls and seeing if the existing system can be used, and if so, would it be less expensive than the project covering the cost of new infrastructure setup? Another option would be to admit the mistake to the grant project officer and ask if cutting scope elsewhere on the project is possible to cover the overage for the lack of security budgeting.

14. What should have happened in the scenario in the two questions above?

    Correct answer: An IT or security professional should have been approached in the planning stages of the project, preferably one familiar with existing institutional infrastructure. The IT professional and the investigator should have searched together for the most cost-effective solution that met the security needs of the project, especially solutions that leverage existing infrastructure.

15. A researcher on a project requires a statistical analysis that he or she is unfamiliar with. He or she is working with a professional statistician. How can the researcher increase his or her confidence that the analysis is correct?

    a. Read up on the methodology and replicate the analysis himself or herself
    b. Ask another statistician to review the tables and figures
    c. Talk with the statistician and plan for an independent analysis
    d. None of the above; if the researcher cannot do the analysis, he or she should not conduct the study

    Correct answer: (c) Talk with the statistician and plan for an independent analysis. Option (c) is respectful of the statistician and will serve to increase the rigor of the research. Any research professional will understand the researcher's need for reproducibility, quality assurance, and due diligence. Ideally, this plan would be discussed in the planning stages and in initial discussions with the statistician, so that the independence of the second analysis would be protected.

References

Barrett S. Why the NIH study of chelation therapy should have been stopped. Chelation watch website. Last updated November 16, 2012. Available from http://www.chelationwatch.org/research/tact.shtml. (Accessed May 7, 2016).

Kolowich S. Security Hacks. Inside Higher Ed. January 27, 2011. Available from https://www.insidehighered.com/news/2011/01/27/unc_case_highlights_debate_about_data_security_and_accountability_for_hacks. (Accessed May 5, 2016).

UNC-Chapel Hill, Professor Yankaskas reach settlement. April 15, 2011 University of North Carolina at Chapel Hill News Archive. Available from http://unc-newsarchive.unc.edu/2011/04/15/unc-chapel-hill-professor-yankaskas-reach-settlement-2/. (accessed May 5, 2016).

Weiss R and Nakashima E. Stolen NIH Laptop Held Social Security Numbers. Washington Post, April 10, 2008.

The Federal Information Security Management Act of 2002 (FISMA), 44 U.S.C., Sec. 3542.

U.S. Department of Commerce and the National Institute of Standards and Technology, Guide for Developing Security Plans for Federal Information Systems NIST-800-18 revision 1, 2006. (Accessed May 1, 2016).

U.S. Department of Commerce, Technology Administration, and the National Institute of Standards and Technology, Federal Information Processing Standards Publication 199 (FIPS PUB 199) Standards for Security Categorization of Federal Information and Information Systems, February 2004. http://csrc.nist.gov/publications/fips/fips199/FIPS-PUB-199-final.pdf. (Accessed May 1, 2016).

Williams M, Bagwell J, and Zozus MN. Data Management Plans, the Missing Perspective. *Journal of Biomedical Informatics*, 2017, in review.

16

Data Ownership, Stewardship, and Sharing

Introduction

Questions of data ownership, stewardship, and sharing have become prominent in research endeavors. Most institutions have policies governing data ownership, research documentation, and records retention. The two major federal funders of basic scientific- and health-related research in the United States, the National Science Foundation (NSF) and the National Institutes of Health (NIH), both require plans for data sharing. This chapter discusses federal requirements and common institutional practices regarding data ownership, management, and sharing.

Topics

- Institutional policies for research data
- Data ownership and stewardship
- Federal requirements for data sharing
- Effective data sharing
- Special considerations for sharing human subjects data
- Special considerations for sharing data received from others

Institutional Policies for Research Data

Most institutions have policies governing the aspects of research data, for example, research data ownership, records retention requirements, and necessary research documentation. Common topics included in institutional policies for research data (Table 16.1) are guided by federal regulation,

TABLE 16.1

Common Aspects of Research Data Governed by Institutional Policy

Institutional definition of data	Data archival requirements
Data ownership	Data access rights
Data stewardship	Data sharing
Essential documentation	Data disposal
Records retention period	

risk management, and patent law. However, topics covered by policy differ from institution to institution on major issues such as what constitutes data, who owns data, and for how long data must be retained. Institutional policies for research data at academic institutions are often located in faculty handbooks and at other organizations with policy or procedural information.

Data Ownership and Stewardship

The Bayh–Dole Act of 1980, also called the University and Small Business Patent Procedures Act, gives U.S. universities, small businesses and nonprofits intellectual property control of inventions that result from federally funded research. This includes data produced from federally funded research. Thus, the act confers ownership of data to the institution. Some institutions maintain that the institution owns the data and that the researcher is responsible for stewardship of the data, whereas other institutions grant data ownership rights to researchers. Thus, if an investigator moves to a different institution, the investigator may or may not be able to take data from his or her research with them.

In research conducted at academic institutions and funded through industry or foundation contracts, the contract usually designates ownership of the data. In a contract situation, it is up to the investigator to assure that the rights to retain a copy of the data, to use the data, and to publish manuscripts based on the data are retained; otherwise the academic nature of the research is in jeopardy. Alternately, research conducted within private and publically traded companies is usually considered *work for hire*, proprietary, and as belonging to the company rather than the researcher. In this case, the company usually requires organizational approval for any release of research data or results.

Regardless of whether or not the researcher *owns* the data, the researcher is usually considered the responsible steward of the data. *Stewardship* means that the researcher is held accountable for adhering to institutional policies where they exist for proper documentation, management, retention, and archival of data. Although methods for these have been covered in Chapters 1 and 15, the assignment of responsibility to the researcher through institutional policy regardless of data ownership requires mention. Although institutions are increasingly requiring such accountability, few provide resources for these tasks and instead rely on the investigator to accomplish them through the funding that supported the research.

Federal Requirements for Data Sharing

Although federal agencies have long held the philosophy that data collected with public funds belong to the public (National science foundation 2005, National Institutes of Health 2003), this principle has only recently been implemented in enforceable ways. The two major federal funders of basic scientific- and health-related research in the United States, the NSF and the NIH, require *plans for data sharing* (National Institutes of Health 2003, National science foundation 2010). In 2010, the NSF started requiring submission of data management plans with grant applications. Each of the seven NSF directorates now has directorate-specific requirements for inclusion of a two-page data management plan with grant applications. The data management plan is subject to peer review with the application, and adherence to the plan is assessed at interim and final project reporting. NSF data management plans pertain more to preservation and sharing of data and other research products than the actual handling of data during the research and include the following information (National science foundation 2010):

- The types of data and other research products, for example, samples, physical collections, software, and curriculum materials to be produced in the course of the project
- The standards to be used for data and metadata format and content
- Policies for access and sharing including necessary protections
- Policies and provisions for reuse, redistribution, and the production of derivatives
- Plans for archiving and preservation of access to research products

Similarly, the NIH published its statement on sharing research data in February 2003 as an extension of the existing policy on sharing research resources (National Institutes of Health 2003). According to the NIH data-sharing policy, "applicants seeking $500,000 or more in direct costs in any year of the proposed research are expected to include a plan for sharing final research data for research purposes, or state why data sharing is not possible" (National Institutes of Health 2003).

Effective Data Sharing

It is rarely sufficient to share just the data. Effective sharing is sharing that facilitates reuse of data by others. Thus, sharing data, and other research resources for that matter, necessitates inclusion of additional documentation defining and describing the shared data. Sharing, as pointed out by the NSF (National science foundation 2010), also includes formatting data using applicable data standards. Use of data standards facilitates combining data from multiple studies, enables use of existing software and computer programs that leverage the standards, and decreases the documentation burden on the data provider and the learning curve of the data recipient.

The National Institute on Drug Abuse Treatment has maintained a *public data share* since 2006 for data from all clinical trials conducted by their network (https://datashare.nida.nih.gov/). For each study, the data share includes (1) a copy of the data collection form; (2) definition and format of each field on the data collection form; (3) the research protocol describing the detailed design, subject selection, conduct, and analysis of the study; and (4) the study data formatted using an applicable international data standard. For clinical trial data, this information is deemed necessary and sufficient for complete understanding of the shared data. In general, shared data should be accompanied by documentation sufficient for someone not involved with the original study to understand and use the data.

Special Considerations for Sharing Data about Human Subjects

The Department of Health and Human Services (DHHS) in Title 45 CFR Part 46, Protection of Human Subjects also called the common rule, defines *human subjects research* as research involving a living individual about whom an investigator conducting research obtains (1) data through intervention or interaction with the individual, (2) identifiable private information, or (3) identifiable specimens.

Institutions engaged in human subjects research, that is, those accepting federal funding for human subjects research, rely on procedures established by the local ethics review board, called an institutional review board (IRB) in the United States, for collection and use of human subjects data. Use of patient data in research has special requirements such as patient consent and authorization, use of a limited data set (LDS) that cannot be relinked to the patients, or use of deidentified data. In the United States, all of these require IRB approval if the study is subject to

U.S. Food and Drug Administration regulations 21CFR Parts 50 and 56, or DHHS regulation 45 CFR part 46.

At institutions engaged in federally funded human subjects research in the United States, there is a requirement for Federal Wide Assurance that all studies conducted at that institution are provided oversight by the IRB. At these institutions, IRB approval often constitutes approval to access and use the data, whereas at other institutions, further departmental or other organizational approval may also be required. There are studies conducted by institutions that do not fall into categories requiring IRB existence and research oversight; however, publication of research involving human subjects now generally requires that the study be reviewed and approved by an IRB. Thus, investigators are advised to locate an appropriate IRB prior to embarking on research using patient data even if their institution may not require it.

The Health Insurance Portability and Accountability Act (HIPAA) privacy rule (45 CFR Part 160 and Subparts A and E of Part 164) requires that covered entities, that is, health-care providers and facilities, health-care plans and clearinghouses, and their business associates, only release protected health information, that is, individually identifiable health information abbreviated PHI, in certain controlled situations. These controlled situations include health-care reimbursement or operations, release to the individual, release to regulatory authorities, release in 12 national priority situations, release with authorization from the individual, and release as an LDS. Such release is subject to the principle of minimum necessary information. Thus, even though the HIPAA privacy rule applies to covered entities, it impacts research use of protected health information through limiting the release of data to two mechanisms: (1) release from each individual patient, called HIPAA authorization, or (2) release of an LDS. An LDS is a set of identifiable information for which many but not all of the 18 HIPAA identifiers have been removed. LDSs are released for research, public health, or health-care operations in which the recipient agrees to restrictions such as securing the data and not reidentifying the information. Organizations require recipients of data to enter into Data Use Agreements (DUAs). Thus, use of health-care data from organizations requires contractual agreement with the organization as well as HIPAA compliance with respect to use and disclosure of the data.

Within the limits of these regulations, the NIH expects sharing of data from human subjects research stating that "Data should be made as widely and freely available as possible while safeguarding the privacy of participants, and protecting confidential and proprietary data" (National Institutes of Health 2003). Similarly, the Institute of Medicine echoed the need to share clinical research data in a recent report (IOM 2013). Safeguarding the privacy and protecting the confidentiality of human subjects data while sharing data include (1) ethics review such as that provided by an IRB, (2) obtaining prior consent and HIPAA authorization for such sharing, and (3) deidentifying or anonymizing data unless sharing is authorized. Thus, even though special protections are certainly required, data from human subjects research are subject to data sharing expectations.

Special Considerations for Sharing Data Received from Others

Today through data sharing, many scientific questions are answered using data collected by others. Use of existing data is called *secondary data use* or *secondary analysis*. Some data are available publicly or through *data sharing plans* as described above, whereas other data are obtained through learning of an existing data set and contacting the original researcher to request access. In these cases, a Data Use Agreement (DUA) is often written between the data provider and the data receiver.

When approaching a researcher for a DUA, an investigator should be prepared to provide the following information:

- Detailed and precise statement of exactly what data elements are required, from what sources, and over what time period
- Description of how the data will be used, for example, the planned research questions
- Description of how the data will be transferred securely to the investigator if required
- List of all who will be permitted to access and use the data, including intentions for further data sharing
- Agreement that the research will not further disclose the information, expect as permitted by the DUA or as permitted by law
- Agreement that the researcher will report any unauthorized use or disclosures to the data provider
- Agreement to safeguard the information, description of security measures, and how and where data will be stored and protected
- Agreement that anyone to whom the data recipient provides the data, for example, a subcontractor or research collaborator, will be held to the restrictions of the DUA
- Agreement not to contact or reidentify the individuals, including agreement not to link the data with other data sets that may enable such reidentification

Thus, the DUA dictates what, if any, sharing of the data by the data recipient is permitted. Any restrictions on sharing secondary use data should be described fully in the data sharing plan of a grant application.

Summary

The chapter covered data ownership and emphasized that institutional policies vary. Often, institutions rather than individual investigators own data, and investigators are assigned the role of steward of the research data for studies that they design and conduct. The chapter discussed the predominant philosophy that data collected with public funds belong to the public and should be made as broadly available as possible while protecting private, confidential, and proprietary data. Sharing data and research resources was also covered as was special consideration with respect to confidentiality and privacy protection for human subjects research.

Exercises

1. Which of the following are possible ways to share data from a study?
   a. By uploading data and documentation to a discipline-specific public database
   b. By noting in the primary publication of the research results that data may be obtained by contacting the author
   c. By including data as supplemental material in the journal article in which the primary research results are published
   d. By publishing a database manuscript in a journal that supports such publications
   e. All of the above

   Correct answer: (e) All of the above.
2. The principal investigator of a study usually owns the data from the study. True or false?

   Correct answer: False. Most often the institution employing the principal investigator owns the data if the ownership is not retained or assigned elsewhere by contract.
3. When an investigator leaves an institution, he or she is free to take the data with him or her without any additional approval. True or false?

   Correct answer: False. Because most often the institution employing the principal investigator owns the data if the ownership is not retained or assigned elsewhere by contract. An investigator usually must get permission to take a copy of the data.

4. Data should never be disposed of. True or false?

   Correct answer: False. The useful life of the data should be determined. Data reaching the end of their useful life should be disposed.

5. If an investigator has DUAs giving him or her permission to use the data covered by the agreement, he or she can further share the data. True or false?

   Correct answer: It depends on what the sharing restrictions are stated in the DUA. Usually, further sharing is not permitted.

References

National science foundation, Long-Lived digital data collections: enabling research and education in the 21st century. September 2005. Available from http://www.nsf.gov/geo/geo-data-policies/nsb-0540-1.pdf. (accessed January 25, 2015).

National Institutes of Health, NIH Data Sharing Policy and Implementation Guidance, March 5, 2003. Available from http://www.grants.nih.gov/grants/policy/data_sharing/data_sharing_guidance.htm. (accessed January 25, 2015).

National Science Foundation, Proposal and award policies ad procedures guide: Part II - Award & Administration Guide, Chapter VI.D.4. October 2010, Effective January 18, 2011 NSF 11-1, OMB Control number: 3145-0058.

National Institutes of Health (NIH), Final NIH statement on sharing research data. Notice: NOT-OD-03-032, February 26, 2003.

IOM (Institute of Medicine), *Sharing clinical research data: Workshop summary.* Washington, DC: The National Academies Press, 2013.

17

Data Archival

Introduction

Data that have been used in an analysis require archival. In some cases, even data that were collected but never used require archival. Thus, most research projects should have an archival component. While it is tempting to relegate archival planning till after a study has concluded, thinking about archival early when writing the data management plan can vastly reduce the time and effort that investigators and research teams put into archival.

Institutions, research sponsors, and regulatory agencies often have different requirements for archival and records retention. This chapter provides a broad overview covering research records, records retention, and records retrieval as well as planning and formatting data for archival. This chapter is not intended to serve as a detailed resource for curation and management of research data repositories; good books exist on the topic and should be consulted. This chapter serves as an overview to create awareness of important archival principles. Different models for archival are presented, and pointers to relevant resources are provided further preparing the reader to plan and coordinate data archival for research projects.

Topics

- Archival and reuse of research data
- What to archive
- Formats for archival
- Records retention
- Discoverability of archived data and research records

Archiving and Reuse of Research Data

Research data and associated documentation are archived in case they are needed. Thus, the purpose of archival is reuse. There are two basic reasons for reuse of data from a study: (1) to answer questions about the research and (2) to answer new research questions.

Whether questions about the research come from readers of published research results seeking clarification, desiring additional information, or questioning aspects of study design or conduct, the archived data and associated documentation serve one purpose—to enable recreation of what was done. Internal and external auditors and regulatory authorities have a similar purpose—to trace data from the final analysis back to the original raw data. This traceability demonstrates that data were collected and handled in accordance with study procedures and increases confidence in the legitimacy of data. In the case of such an inspection, documented procedures are required and serve as a basis for the inspection. Another commonly used basis is applicable regulations. Thus, having well described, tested, and maintained, procedures that are compliant with applicable regulations is the first step to a successful outcome. The second step is having sufficient artifacts or evidence that demonstrates faithful execution of the procedures. Such evidence often includes date and time stamps of individual data transactions, documentation of changes to data, version histories of written documents and computer code, log files from execution of computer programs, and documentation of *exceptions*—instances where something odd happened, and a procedure was not or could not be followed as intended. The third step to a successful audit is that the procedures (step 1) and the artifacts of their execution (step 2) match, that is, that the artifacts demonstrate faithful compliance with written procedures. Audits of research are often performed by institutions themselves, by funders of research, and by regulatory authorities. Many institutions have internal auditing functions that routinely audit projects as part of an institutional quality assurance or compliance program. Writing procedures such that they are auditable, i.e., providing for evidence that the procedure was followed (or not), helps the auditing process run smoothly.

This very simple concept, demonstrating that procedures were followed, when applied to research projects, often generates pages of inconsistencies also called *nonconformities*. Inconsistencies or nonconformities are objective facts about instances where step 1 and step 2 did not match. At the end of an *audit*, an auditor will group similar inconsistencies together into observations. For example, if all changes to data in laboratory notebooks were to be dated and initialed according to the data recording procedure, and an auditor found 20 instances of this not being done, the observation would read something as follows: *"changes to data were not documented" as required by the Standard Operating Procedure (SOP)*, or, *"lack of oversight regarding implementation of the data recording SOP"*. These observations would be supported by the list of the inconsistencies. If, in addition, the dating and initialing of changes were performed inconsistently between members of the laboratory, the observation might read something as follows: *"inconsistency in following the data recording SOP and lack of oversight of laboratory personnel in regards to compliance with the SOP"*. This type of comparison is often done during a process audit.

As an aside, Stan Woolen, former deputy director of the U.S. Food and Drug Administration (FDA), is reported to have said "May the source be with you" in reference to the occurrence of a U.S. FDA audit or inspection. The FDA also expects data in the source to be traceable all the way through to the analysis (Woollen 2010) as do consumers of published research who expect researchers to be able to answer questions about a study sometimes long after it has been published. Therefore, in the case of an audit or a simple question,

> may the source be with you;
> may it be appropriately reflected in the analysis,
> may the path from the source to the analysis be clear,
> and may it demonstrated by the study archive!

Although audits can (and often are in best practice organizations) be done during a study, they are often conducted after study closure and depend on archived data. Such an audit is much easier to prepare for, if archival planning and preparation have been done on an ongoing basis as part of maintaining good study documentation. The same combination of well-described procedures and artifacts that demonstrate their faithful execution also supports a researcher in looking back and answering questions about the research that come up from readers or others.

Answering questions about published research relies on traceability—what was done, when, by whom, and how it is reflected in the data, associated documentation and the analysis. The extreme case of answering questions about the study is independent replication of the study itself. *Reproduction*, however, can come in many forms of varying degrees of independence:

- Reanalysis of the data by an independent statistician associated with the laboratory
- Reanalysis of the data by an independent group
- Collection and analysis of new data following the same protocols

For all but a completely independent replication, documentation of study design and conduct is critical in performing a high-fidelity reproduction. In the case of a failed replication or reproduction, the number of which has been shown recently to be high (Freedman 2015, Open Science Collaborative 2012, Baker 2015, Morrison 2014), the details of how data were collected, handled, and analyzed become paramount in identifying differences between the study and the failed replication or reproduction. These explorations are critical not only in testing study reproducibility but in the new information we learn from them.

The second basic reason for which data and associated documentation from a study are reused is to answer new questions. Answering new questions requires good documentation about data definition and about how data were processed. Consider a geological survey conducted 10 years ago in six sites. A second researcher comes along and is interested in validating a new methodology by comparing his or her data to the previous study. He or she acquires the data from the first researcher and picks six sites within the same 100-mile radius area. When he or she compares the results, he or she gets markedly different answers. He or she goes back to the first researcher and finds that the initial study was done in six sites in old and now dry riverbeds, and that the first survey used small bore hand augering to a depth of 2 feet rather than his large bore drilling to a death of 10 feet. The initial data set did not document the site selection, augering method, or bore diameter. As can be seen, differences in study design and conduct can render data useless for many secondary uses and complete documentation of the initial study is required to assess appropriateness of existing data for new uses.

Documentation at the data element level and operational definitions about how data were processed are equally important. For example, consider a psychosocial study in which the data are publically shared, but the study start dates have been randomly offset in month and day with a ±2 year shift, and all other study dates offset to maintain the day number intervals. A second researcher would like to use the data to study seasonal variation in the data. Knowing how the data were processed, in particular deidentified and date shifted for public sharing, prevented what could have been an unfortunate misuse of the data. Reuse of data to answer new questions is less about traceability and more about the definition of the study population, the definition of the data elements, and how the data were collected and processed that may impact how the data can be interpreted or used.

What to Archive

Chapter 4 discussed data management plans. Recall that the data management plan is written before the study starts and is updated as needed while the study is ongoing. Also during the conduct of a study, the operations associated with collecting and processing the data generate by-products or artifacts that evidence how data were processed. An *operation* is a single action on a data value. For example, entering a data value into a database is an operation. A separate operation would be conversion of a data value to a different unit of measure. Upon entry, many systems also record the user identifier (the user ID) of the account entering the value and the date and time of the entry (the time stamp). The entered value, user ID, and timestamp are artifacts left by the operation documenting what occurred (the entered value), when it occurred (the timestamp), and who or what made the transaction (the user ID). There are often as many artifacts as data values for a research study in which a system with an audit trail is used. Not all artifacts are technical controls, i.e., automated and electronic, like audit trail artifacts. Artifacts can also be signatures or documents. For example, if the data management plan for a study stipulates that the data will be audited at the midpoint of the study and at the end, and that all identified data discrepancies will be corrected and that the error rate will be calculated, the expected *artifacts* include the *audit report* containing an error rate and some record of the individual data discrepancies identified and their disposition, that is, their resolution. Another type of artifact is an exception or incident report. There are times when procedures cannot be followed, for example, if an instrument fails and a measurement is missed or taken 10 minutes late because it took time to power and start the backup instrument. Exceptions such as these are documented in a brief *exception report* that states why a procedure could not be followed in the particular instance and how the situation was handled with particular attention to human safety or animal welfare if pertinent and handling of data. Artifacts can be as simple as training records demonstrating that required training occurred and signature approval of updated SOPs signifying that the revision and approval process were followed. All of these artifacts demonstrate that procedures were faithfully executed and as such become part of the study documentation, sometimes also called *the study record*. Thus, as a procedure specifies that something should be done, ideally there is an artifact documenting the performance of the procedure. The presence of an artifact is also referred to as *auditable*, meaning that the fact that something occurred and information about it can be confirmed (or refuted). Operations that are documented and auditable, are traceable. Records of all operations are part of the study record and should be archived with the study.

The information that should be archived with a study includes the data management plan components and the associated artifacts (Table 17.1). The items in the leftmost column reflect data management plan components from Chapter 4. Each row contains an item covered elsewhere in this text. The center column describes artifacts that are generated during the conduct of the study. The right most column follows through to list the items that should be archived with the study data. Archival of the complete items in Table 17.1 will enable a study to be reproduced and will contain all of the information that someone using the data would need.

TABLE 17.1

Documenting Data

Before the Study	During Study Conduct (Plan Updates, Exceptions and Artifacts)	After the Study (Shared or Archived Data and Documentation)
Study design or research protocol and associated study conduct materials, for example, forms and job aids	Documented changes	Study design or research protocol and all amendments
Planned population description	Documented exceptions	Population characterization
Investigator and research team member role, responsibilities, duration on the project, data access, signature and identity log, and demonstration of training and other qualifications	Documented changes	Record of all research team members, their role(s), duration on the study, system identifiers (user IDs), curriculum vitae or resumes and training records
Description of planned data sources and definition of data elements	Documented changes to data sources or data element definition	Definition of data elements in data set and record of changes to data definition
Documentation of data flow and workflow	Current data flow and workflow diagrams Record of changes to data flow and workflow diagrams	Description of workflow and data flow for the study Changes and exceptions to workflow and data flow
Planned data models used to store and share data	Documented changes to data models	Data model and table specifications for active- and inactive- phase datamodels used for the project. If any of the specifications are written against a data model, the data model should be included
Procedures for measurement or observation	Documented exceptions and changes to procedures	Final procedures and record of exceptions and changes to procedures
Instrumentation used and procedures for calibration and maintenance	Documented exceptions and changes to instrumentation or procedures; instrumentation calibration and maintenance records	Final procedures and record of exceptions and changes to them Instrumentation calibration and maintenance records

(*Continued*)

TABLE 17.1 (*Continued*)

Documenting Data

Before the Study	During Study Conduct (Plan Updates, Exceptions and Artifacts)	After the Study (Shared or Archived Data and Documentation)
Procedures for recording data including forms, field or laboratory notebook templates, or entry screens	Documented exceptions and changes to procedures and forms for recording data Attribution and date and time stamp of original data entries	Final procedures and forms and record of exceptions or changes to them. Raw data as originally recorded Attribution and date and time stamp of original data entries
Procedures for data processing including cleaning, imputations, standardization and other transformation	Documented exceptions and changes to procedures	Final procedures and record of exceptions or changes to them Record of all changes to data values
Procedures for transfer and integration of external data	Documented exceptions and changes to procedures. Data transfer, receipt, load, and exception and integration logs	Final procedures and record of exceptions or changes to them Data transfer, receipt, load, and exception and integration, logs
Procedures for identification and handling of discrepant data (if not described above as data cleaning)	Documented exceptions and changes to procedures. Discrepancy tracking and disposition. Record of all data updates resulting from discrepancy resolution	Final procedures and record of exceptions or changes to them; record of all discrepancies, their disposition, and changes to data
Description of systems in which data are to be stored including description of major system functionality used as well as plans for scheduled and unscheduled system down-time	Specifications for computer programs and project-specific configuration, including changes All versions of source code that operated on data; specifications for changes. Documented exceptions and changes to data storage plans	Description of systems in which data were stored. Documentation of computer system validation or copy of authorization to operate. May consider archiving the computer system itself if data archival format is proprietary. All versions of source code that operated on data; specifications for changes.
Software validation and change control plans	Validation or testing plans commensurate with risk Documented exceptions and changes to data storage plans	Testing documentation commiserates with risk. Documentation of successful testing. Final procedures and record of exceptions or changes to them

(*Continued*)

TABLE 17.1 (*Continued*)

Documenting Data

Before the Study	During Study Conduct (Plan Updates, Exceptions and Artifacts)	After the Study (Shared or Archived Data and Documentation)
Data security plan (computer systems and noncomputerized data and study documentation)	Changes to data security plan Incident reports for security breaches	Final procedures and record of exceptions or changes to them System monitoring logs or log review records Summary of data security including incident reports
Data backup plan and schedule including location of stored backups and duration for which they are maintained	Changes to data backup plan and schedule	Final procedures and record of exceptions or changes to them System monitoring logs or log review records Summary of backups including incident reports
Planned deliverables or milestones and associated dates; staffing projections and other management tools used such as data status reports	Changes to deliverables, milestones, or staffing. Minutes from meetings where data-related decisions were made; copies of communications documenting data-related decisions	Deliverable and milestone dates. Description of scope changes and documentation of staffing changes. Minutes from meetings where data-related decisions were made; copies of communications documenting data-related decisions
Quality assurance and control procedures	Documented exceptions and changes to procedures. Quality management system maintenance artifacts	Final procedures and record of exceptions or changes to them. Copies of quality management system artifacts such as SOPs and SOP revisions, position descriptions, resumes/curriculum vitae, and training documentation if not included elsewhere. Quality control records and results of data quality assessments

(Continued)

TABLE 17.1 ***(Continued)***

Documenting Data

Before the Study	During Study Conduct (Plan Updates, Exceptions and Artifacts)	After the Study (Shared or Archived Data and Documentation)
Privacy and confidentiality procedures and data to which they apply	Documented exceptions and changes to procedures	Final procedures and record of exceptions or changes to them
Description of procedures for data sharing, archival, maintained retention, and disposal	Documented exceptions and changes to procedures	Final procedures and record of changes to them
	Data and description of archived or shared format	Copies of study data used for study decision making, for example, stopping rules or interim analyses and for the data on which the final analysis are based Description of the format in which data are archived

Formats for Archival

Careful thought should be given to the format in which data are shared or archived. Archiving data in proprietary data formats makes them less accessible in the future, because proprietary formats usually require the software that generated the data to read or display the data. Thus, proprietary formats should be avoided whenever possible for archiving. Two formats that have thus far stood the test of time are delimited text files using the ACSII character set and Extensible Markup Language.

Data are often shared using *common data models*. These exist for many disciplines. Archiving data in a common data model has advantages where a secondary user of the data also has data in that common data model and wants to combine the two data sets; having both in the common data model will save time and will ease the burden of learning the shared or archived data. Further, common data models often provide some level of standardization of data definition either by virtue of where the data are stored in the model or by accompanying information. The disadvantage of using a common data model, if it is not an exact match to the one that used to collect and manage the data, is that the study data have to be mapped to the data model which takes effort. Additionally, such mapping usually results in some *information reduction* and loss. Thus, use of common data models should be considered and decided in the early planning stages of a study.

Records Retention

Results and data submitted to regulatory authorities often have required records retention periods. Academic institutions and companies also often have required retention periods for research records and data from some or all studies conducted at the institution. These do not always agree; thus, researchers should be familiar with both and should be prepared to use the longer retention period of the ones that are relevant to a project.

Discoverability of Archived Data and Research Records

Data *curation* has been variously defined. Here, data *curation* is the association of data with definitional information and documentation describing the data's origin and handling such that data can be appropriately interpreted and used by someone not involved with the study from which the data were produced.

Those same principles apply to archived data as well. Consider a study that was archived mostly in paper records with a total of 98 banker boxes sent to an institutional storage facility. Each box was numbered and associated with the study. Three years later, a question arose about whose responsibility it was to archive the study and for how long. Ironically, the scope documentation, including a description of responsibility for archival and records retention, was archived with the study. Unfortunately, no one created a manifest indicating what was in each box, and all 98 boxes had to be retrieved and examined to locate the scope documentation. To prevent this scenario, archived materials, whether paper or electronic, require curation and provision for discoverability such that archived information can be located. If for no other reason, careful curation lessens the effort required for eventual disposal. When planning for curation, a good rule of thumb to use is the amount of time that it would take to locate a piece of information at a given level of specificity in curation. If the time estimate for retrieval is too long, given the likelihood that a retrieval need will occur, then additional detail is needed.

Summary

In summary, this chapter covered the reasons for archival of study data and records. The data management plan framework from Chapter 4 was expanded to provide a comprehensive list of information that should be archived with study data to support answering questions about the research and using the study data for new research questions. Consideration for records retention, format of archived data, curation, and discoverability were briefly discussed in Chapter 16.

Exercises

1. Locate your institution's policy on available resources for records retention and archival.
2. Compare and contrast your institution's records retention and archival policy with that found in regulations applicable to your discipline.
3. Locate the records retention period for the most likely funder for your research.
4. Describe how a researcher should approach archival of a project conducted by a postdoc in his or her laboratory.

Correct answer: Table 17.1 could be used as a template. The researcher should discuss with the postdoc what items from the table should be archived, the archival format, and the curation and organization of the archive. The postdoc and research team can execute the archiving process. The researcher should spot-check to make sure that the organization is sufficient and that needed records are retrievable.

5. Locate a shared data in your discipline. Compare and contract the archival and curation with that from Table 17.1.
6. For the shared data in question 5, describe your ability to understand and use the data file to do descriptive statistics characterizing the study population or experimental/observational units. What if any information did you find hard to locate, and which did you find useful?
7. If an organization has SOPs that cover the procedures used for a study, the procedures do not need to be restated in the data management plan. True or false?

   Correct answer: True. The data management plan should describe study-specific aspects that are not otherwise covered by higher level organizational documentation.
8. A small research laboratory that does not have SOPs needs to develop them, so the data management plan can refer to them. True or false?

   Correct answer: False. A research team that does not have SOPs can document the needed procedures in the data management plan.
9. As long as data and data element definitions are archived, a study can be reconstructed. True or false?

   Correct answer: False. The full set of procedures for how the data were collected and processed in addition to the artifacts demonstrating following (or not) of procedures is required.
10. An auditor auditing the conduct of a study is likely to compare which of the following:
    a. Organizational SOPs to relevant regulations or other requirements
    b. Study procedures to relevant regulations or other requirements
    c. Study procedures to study artifacts
    d. All of the above

    Correct answer: (d) All of the above. (a) and (b) reflect the adequacy of the procedures (see Chapter 13) and (c) reflects extent to which the procedures were followed.

References

Woollen SW. Data Quality and the Origin of ALCOA. *The Compass—Summer 2010. Newsletter of the Southern Regional Chapter Society or Quality Assurance.* 2010. Accessed November 17, 2015. Available from http://www.southernsqa.org/newsletters/Summer10.DataQuality.pdf.

Freedman LP, Cockburn IM, and Simcoe TS. The Economics of Reproducibility in Preclinical Research. *PLoS Biol.* 2015; 13(6):e1002165.

Open Science Collaboration. An Open, Large-Scale, Collaborative Effort to Estimate the Reproducibility of Psychological Science. *Perspectives on Psychological Science* 2012; 7(6):657–660. doi:10.1177/1745691612462588.

Morrison SJ. Time to do something about reproducibility. *Elife* 2014; 3. doi:10.7554/eLife.03981.

Baker M. First results from psychology's largest reproducibility test. *Nature News* 2015. doi:10.1038/nature.2015.17433.

Glossary

abstracting data: a process in which a human manually searches through an electronic or paper document or documents (also called the record) that may include structured or narrative data to identify data required for a secondary use. Abstraction involves some direct matching of information found in the record to the data required and commonly includes operations on the data such as categorizing, coding, transforming, interpreting, summarizing, or calculating. The abstraction process results in a summary of information about a patient for a specific secondary data use (Zozus et al. 2015, PLoS One).

acceptance criteria: limits for data quality which if reached requires a specified action to be taken.

accuracy: the state of consistency between a data value and a true value at a stated or implied point in time. Complete accuracy is an identical match. Some values may be deemed consistent enough for a particular data use, even though they are not an exact match.

active phase of a study: the time period during which data are accumulated are being processed and may change.

ALCOA: a mnemonic for attributable, legible, contemporaneous, original, and accurate coined by Dr. Stan Woollen during his tenure at the U.S. Food and Drug Administration.

asynchronous: activities that occur with a delay in time between some initial action and a response to that action.

attributable: a property of a data value. Attribution exists when a data value remains in association with its origin, including the individual who or device which measured, observed, or recorded it.

audit trail: in data systems, a record of all changes to data values since their initial entry into the system. A true audit trail cannot be altered.

automation: performing all or part of a set of tasks with a machine rather than through human effort (NRC 1998).

barcode: digital or printed lines or boxes that encode information and that are read with special software. Reading barcodes printed on paper or labels requires a special device.

bias: in the context of data, bias is the over or underestimation of a parameter due to some systematic and usually unwanted issue in the data. For example, in a study with a high number of lost to follow-up individuals, if a higher number in the experimental arm was due to death secondary to lack of efficacy of the experimental intervention, the estimation of treatment effect obtained from patients who complete the study may be biased.

blind verification: a double data entry process where the second entry operator is not aware of the value entered by the first entry operator.

central tendency: some measure of the typical value of a distribution of data. Example measures of central tendency include the average, the median, and the mode.

check all that apply list: a data collection structure for capturing discrete data for which one or more option is expected from respondents.

check only one list: also called a radio button; a data collection structure for capturing discrete data for which only one option is expected from respondents.

coding: a data standardization process of applying standard codes to data values.

common data model: an agreed upon structure for storing or sharing data, including data standards and referential integrity.

computer system validation: a formal process of confirming that a computer system is installed according to specification and is operating as expected.

conceptual definition: a conceptual definition explains the meaning of the concept or construct, usually describing the general character and key aspects that distinguish it from other related concepts or constructs.

configuration: altering the way that a computer system operates by setting controls provided by the software without having to write computer programs or change the underlying software itself.

confounder: a covariate that is related to both the dependent and independent variables.

constant: a data element for which the value remains the same.

contemporaneous: a property of a data value that exists when the data value was observed and recorded when the phenomena to which the data value pertains occurred.

continuous variable: a variable that can take on any value between two specified values.

control file: a file that contains mapping of data values usually by data element, to the study-specific data storage structures. The control file is the instructions used by the software for importing or loading data into a data system.

covariates: variables other than the dependent and independent variables that do or could affect the response. Covariates are observable.

cumulative data transfer: a data transfer where all of the data are sent, even those that have been previously sent and received.

data: (1) the Diaphoric Definition of Data (DDD) as described by Floridi states that, "a datum is a putative fact regarding some difference or lack of uniformity within some context" (Floridi 2017),

(2) the international standard ISO/IEC 11179 defines data as, "a re-interpretable representation of information in a formalized manner suitable for communication, interpretation, or processing" (ISO/IEC 11179 2003), and (3) DAMA defines data as, "Facts represented as text, numbers, graphics, images, sound, or video…" (Mosley 2008).

data cleaning: the processes of identifying and resolving potential data errors.

data collection: with respect to research, data collection is the recording of data for the purposes of a study. Data collection for a study may or may not be the original recording of the data.

data consistency: a state of agreement or conformance to an expected relationship between two or more data values.

data discrepancy: an inconsistency where one or more data values do not match some expectation.

data element: a data element is the pairing of a concept and a response format. The ISO 11179 standard lists the information required to unambiguously define a data element.

data enhancement: the process of associating data values with other, usually external, data values or knowledge.

data error: an instance of inaccuracy.

data flow: the path, in information systems or otherwise, through which data move during the active phase of a study.

data flow diagram (DFD): a visual depiction using standard symbols and conventions of the sources of, movement of, operations on, and storage of data.

data handling: the activities that include operations performed on data following their origination. Data handling is sometimes used as a synonym for data management.

data integration: merging or otherwise associating data usually from disparate sources but that pertain to the same experimental or observational unit.

data lifecycle: covers the period of time from data origination to the time when data are no longer considered useful or otherwise disposed of. The data lifecycle includes three phases, the origination phase during which data are first collected, the active phase during which data are accumulating and changing, and the inactive phase during which data are no longer expected to accumulate or change, but during which data are maintained for possible use.

data management: (1) data management in a research context is the application of informatics theories and methods to the definition, collection, and processing of data throughout the data lifecycle for data from a study. Research data management includes the design of data flow and workflow to assure availability of data at appropriate quality, time, and cost. The profession promotes confidence in research results by providing data capable of supporting the conclusions drawn from a study. (2) As defined more broadly in an IT context, data management is the "function that develops and executes plans, policies, practices and projects that acquire, control, protect, deliver and enhance the value of data and information" (Mosely 2008).

data management plan (DMP): an *a priori* description of how data will be handled both during and after research. A DMP is maintained and updated throughout a study and documents data produced from a study.

data reconciliation: reconciliation is a process to detect discrepancies and possible errors in combining data from disparate sources that pertain to the same experimental or observational unit.

data sharing: making data available for use by others.

data standardization: a process in information systems where data values for a data element are transformed to a consistent representation.

data stewardship: assigned responsibility for the collection, processing, storage, security, and quality of data.

data transfer: moving data between systems or organizations.

data transformation: an algorithm-driven process in which data are reformatted, restructured, or mapped to some other value set.

database lock: the point in time when write privileges are removed and data can no longer be changed. Database lock is the start of the inactive phase.

data type: a data type is a category of data. There are many different categorizations of data types. Categorizations are often based on the values that a category can take on, some programming language in which the data type is used, or the operations that can be performed on the data type.

dependent variable: the response or outcome variable in a study. A dependent variable is believed to be influenced by some intervention or exposure.

dependent verification: also called unblind verification is a double data entry process where the second entry operator sees the actual value entered by the first entry operator.

derivative data: facts that we calculate or otherwise reason from other data.

deterministic record linkage: joining data from two disparate sources for the same experimental or observational unit using exact match of some identifier.

dichotomous variable: a dichotomous variable is a discrete variable that has two possible values, such as a lab result that is returned as positive or negative.

direct electronic capture of data: also called eSource is the process of obtaining data in electronic format from a measuring device or a system where a form filler entered the data into a computer.

direct system interface: a data transfer that occurs in automated fashion between two or more data systems.

discrete variable: a variable that can only take on a finite number of values, usually whole numbers.

double data entry: also called double entry is a process where data are entered twice, usually by two different individuals from a paper form or other image into a database system, statistical analysis package, or spreadsheet. The three main types of double data entry include double entry with dependent verification, double entry with independent verification, and double data entry with third party compare.

exceptions: in information processing, exceptions are data values or groups of data values that do not conform expectations.

exhaustive: with respect to a set of concepts or a set of responses to a question, an exhaustive set conceptually covers the whole domain of interest leaving no gaps.

extract transform and load (ETL): the process of writing data out of one data system, transforming it in some way and importing it into another data system.

formative assessment: an evaluation performed before a product or service is final for the purpose of improvement.

free text field: a data collection structure often called a write-in field or fill in the blank field that collects unconstrained text responses provided by a form filler.

functionality: the features of a software product including tasks performed by the software.

general purpose data types (GPD): a categorization of data types put forth by the International Organization for Standardization (ISO) ISO/IEC 11404 (ISO/IEC 11404 2007).

hard automation: automation that requires computer programming to be altered if changes are required.

image map: a data collection structure that collects some location or position in physical space by allowing an observer to record where something is located relative to other spatially located items.

importing data: see loading data.

imputation: the process of systematically replacing missing and sometimes outlying values with an estimated value.

inactive phase of a study: the time period during which data are not expected to change. The inactive phase starts with database lock.

incremental data transfer: a data transfer where only new or changed values (including new deletions) are included.

independent variable: a variable believed to influence an outcome measure.

information: the General Definition of Information (GDI) states that information equals data plus meaning and goes on to require that information consists of one or more data, that the data are well formed, and that the well-formed data are meaningful (Floridi 2017).

information loss: the reduction in information content over time.

information degradation: the reduction in information accuracy or completeness over time.

integrity of research operations: the state that exists when (1) the research plan complies with organizational policy and applicable regulations, (2) research procedures comply with the research plan, and (3) the compliance is evidenced by the data, documentation, or other artifacts.

information reduction: decreasing the number of measures, the detail level of measures, or the scale of the measure, for example, discretizing a continuous measure.

Intelligent character recognition (ICR): recognition and translation of hand-printed characters to structured electronic data.

interactive verification: a data entry process where the first and second entry values are compared by the computer system, as data are entered with the second entry operator addressing differences, deciding the correct value, and making corrections during second entry. The two types of interactive verification include blind verification and unblind or dependent verification.

interval scale: according to Steven's scales, the interval scale type is a numerical scale. The interval scale is quantitative, and the exact value is knowable. Interval scales carry additional information content over that of ordinal scales, because the space between points on an interval scale is the same.

keyboard-based data entry: also called key entry is a process where a human data entry operator reads data from paper or an other image and types the data into a computer. The four main types of key entry include single entry, double entry with dependent verification, double entry with independent verification, and double data entry with third party compare.

latent variable: a variable that is not observable but may affect the relationship of the independent and dependent variables.

legible: a property of a data value. A legible data value is a data value that can be read by the intended reader (for example, a machine or a human).

Level of abstraction (LoA): the detailed level at which something is presented. For example, a house can be represented by a square with a triangle on top (a very high level of abstraction) or a set of engineering drawings (a low level of abstraction).

loading data: also called importing data, the process of reading electronic data and writing the data values to specified locations in a data system.

loss of confidentiality: unauthorized disclosure of information (44 U.S.C., Sec. 3542).

loss of integrity: unauthorized modification or destruction of information (44 U.S.C., Sec. 3542).

loss of availability: disruption of access to or use of information or an information system (44 U.S.C., Sec. 3542).

managerial controls: (1) setting performance standards, measuring performance and taking corrective actions when necessary, for example, definition of job responsibilities, reviewing work, and providing feedback. (2) As an information system security control, "safeguards or countermeasures for an information system that focus on the management of risk and the management of information system security" (FIPS 200).

mapping: in information system parlance, the process of associating data values to another set of data values usually but not always using standard codes.

measurement: quantitation of some aspect of a phenomena of interest to a research study, sometimes but not always a physical quantity; a measurement is the origin of a measured data value.

metadata: often referred to as data about data. Metadata can be a definition of a data element or units of measure or can include information about how data were handled or processed.

mutually exclusive: with respect to a set of concepts or a set of responses to a question, mutual exclusivity is achieved when there is no overlap between the response options.

narrative text field: a free text field data collection structure that collects long unconstrained text responses provided by a form filler, usually multiple sentences or paragraphs of text.

nominal scale: according to Steven's scales, a nominal scale is a label, for example, gender, hair color, or country of origin. Items on a nominal scale have no order and do not have any numerical significance.

nonrepudiation: the state achieved when users cannot deny accountability for actions taken under their user account in an information system. Specifics of nonrepudiation can be found in the electronic record and electronic signature rule, Title 21 Code of Federal Regulations Part 11 (FDA 2003).

observation: a notice of some aspect of a phenomena of interest to a research study; an observation is the origin of an observed data value.

obvious corrections: changes to data made without a resolution or confirmation process.

on-screen checks: computer programs that run during data entry to detect and notify a form filler of discrepant or invalid data. On-screen checks may run real time or immediately after an attempt to submit data for one or more screens.

ontology: a multihierarchy system composed of multiple types of relationships between terms where the terms are defined logically by their position in the hierarchies and relationship to other members of the ontology.

operational data: data that are generated during the collection, processing, and storage of data. A date and time stamp applied when data are first entered into a system is an example of operational data.

operational definition: in research, an operational definition explains how the concept or construct will be measured.

optical character recognition (OCR): recognition and translation of machine print or typed information to structured electronic data.

optical mark recognition (OMR): recognition and translation of marks on paper or other images such as filled-in *bubbles* or ticked boxes to structured electronic data.

ordinal scale: according to Steven's scales, an ordinal scale consists of labels, and additional information through some inherent order, for example, small, medium, large or mild, moderate, and severe. Concepts measured on an ordinal scale are qualitative and without numerical significance.

origin of the data: occurrence of the measured, observed, or asked phenomena; the occurrence of the phenomena or event that produced the data.

original observations: original observations are raw data, that is, facts we observe, process, and store about things in the real world.

original recording of data: also called the source (ICH E6 GCP) is the first time that a data value is recorded on any media.

outlier: a data value that is unlikely given the distribution of the rest.

planning phase of a study: the period of time from project conception to the start of data collection.

policy: a written statement by an organization of what will be done in certain situations.

preprocessing: activities that must be done prior to some downstream processing. Usually but not always, these activities check for things that would hinder or altogether preclude further processing. Examples of common preprocessing activities include checking that all expected pages are present for multipage forms and checking to see that all pages (or items otherwise received separately) are appropriately labeled with the identifier for the experimental or observational unit and time point.

polychotomous variable: a discrete variable having more than two valid values.

preload data checks: checks for nonconformant, invalid, errant, or discrepant data performed prior to importing data into a data system.

primary data use: the original purpose for which the data were collected.

procedures: written descriptions of work processes; descriptions of the steps by which work is accomplished.

prolectively recorded data: data recorded after initiation of the study.

prospective study: a study that looks forward in time from an event of interest, for example, an exposure or some experimental intervention toward some anticipated outcome or lack thereof.

procedural controls: (1) documented processes for doing things. Procedural controls are often written and referred to as protocols, manuals, or standard operating procedures. (2) Also called operational controls; as a security control, "safeguards or countermeasures for an information system that are primarily implemented and executed by people (as opposed to systems)" (FIPS 200).

quality assurance: planned and systematic activities, such as standardizing processes, hiring qualified staff, and providing appropriate training, for the purpose of providing confidence that the desired quality will be achieved (ASQ entry for quality assurance/quality control 2016).

quality control: quality control on the other hand is generally thought of as ongoing operational activities to measure and manage quality. Quality control occurs during a process or study and serves as a feedback loop testing and reporting the quality, so that corrections can be made when needed.

quality data: "Data strong enough to support conclusions and interpretations equivalent to those derived from error-free data" (IOM 1999).

quality management system (QMS): "a framework for planning, executing, monitoring, and improving the performance of quality management activities" (ISO 9000:2015).

radio button: see check only one list.

ratio scale: according to Steven's scales, a ratio scale is numeric and carries the maximum amount of information. Not only is the distance between two points on the scale the same, but there is an absolute zero, for example, temperature measured in degrees Kelvin, weight, or height.

reliability: is traditionally defined as the degree to which repeated measurements of the same thing return the same value.

representational inadequacy: a property of a data element indicating the degree to which a data element differs from the desired concept.

research replication: obtaining equivalent results of a scientific study by independent investigators, using independent data, methods, equipment, and protocols (Peng R, 2009).

research reproducibility: obtaining the same results as the original scientist starting from the data gathered by the original study.

retrolectively recorded data: data recorded before initiation of the study.

retrospective study: a study that looks backward in time from some event of interest (an outcome), for example, to see if some past exposure is associated with the outcome.

rules of behavior: in information security, a set of rules, norms, proper practices, or responsibilities for individuals interacting with data.

run chart: a quality control chart typically used in statistical process control to indicate if a process is in a state of consistent operation that is free from nonnatural variation.

sampling rate: the number of samples taken per unit of time.

secondary use of data: any use of data other than that for which the data were originally collected. Also referred to as reuse of data.
secondary use of data: any use of data other than that for which the data were originally collected. Also referred to as reuse of data.
security controls: the "management, operational, and technical controls (i.e., safeguards or countermeasures) prescribed for an information system to protect the confidentiality, integrity, and availability of the system, and its information" (FIPS 199).

semiotic triangle: an informational graphic showing relationships between things in the real world, mental concepts of them, and symbols used to represent them. The ideas behind the semiotic triangle are credited to Charles Sanders Peirce, an American philosopher and scientist (1839–1914).

separation of duties: also called segregation of duties, is an internal control where more than one person is required to complete a task for the purpose of preventing fraud or error.

single entry: a process where data are key entered one time from a paper form or other image into a database system, statistical analysis package, or spreadsheet.

speech-to-text (STT) processing: recognition of spoken language and translation to written words usually in word processing software, text files, or some display format. STT may also be called automatic speech recognition or voice recognition.

soft automation: automation that is configurable through software without requiring changes to the underlying code of the software itself.

soft delete: inactivation of a data value such that it does not show on the screen or in reports, but the data system retains a record that the data value once existed.

Stevens' scales: are four exhaustive and mutually exclusive categories, nominal, ordinal, interval and ratio, based on the information content of measurement.

structured document: a collection of usually narrative data formatted in a consistent way where the consistency is strict enough, so that computers can process the document sections.

structured fill in the blank field: a data collection structure that holds free text within the bounds of some structure, for example, a two-numeric character day, a three-character month, and a four-numeric character year where the day must be between 1 and 31, and the month must be in the set (Jan, Feb, Mar, Apr, May, Jun, Jul, Aug, Sep, Oct, Nov, and Dec), and the year must fall within some applicable range.

study record: artifacts demonstrating planned procedures and their faithful execution.

summative assessment: an evaluation performed after a product or service is final to provide a measure or overall judgment of quality or compliance.

surrogate: with respect to data, a surrogate measure is something that is measured or observed in lieu of the desired thing, usually because the desired concept cannot be directly measured or is not practical to measure or otherwise observe. A good surrogate measure has been validated to correlate strongly with the desired concept.

synchronous: activities that occur in real time such that there is no appreciable time delay between the initial action and a response.

taxonomy: a single hierarchy built from one relationship, such as is_a, or type_of.

technical controls: (1) automated constraints that assure compliance with a process and (2) as a security control, constraints from an "information system that are primarily implemented and executed by the information system through mechanisms contained in the hardware, software, or firmware components of the system" (FIPS 200).

third person compare: the data from first and second entry are compared usually electronically, and a third person, different from the first and second entry operators, reviews the list of differences, decides the correct value, and makes corrections.

traceability: the state achieved when raw data can be reconstructed from the file or files used for the analysis and study documentation and vice versa.

transcribing: the act of copying or reformatting data form one format to another.

unit conversion: a calculation that changes the representation but not the actual quantity represented or the dimensionality of a data value.

validity: the state of consistency between a measurement and the concept that a researcher intended to measure.

variable: a data element for which the values are expected to change over time or individuals.

visual analog scale: a graphic data collection structure for a continuous variable. The data are collected as the measurement of the distance from an anchor (one end of the scale) to the respondent's mark.

vocabulary: also called a dictionary. A list of terms and their definitions.

workflow: the tasks and the sequence of tasks performed by humans and machines in a process.

workflow diagram: also called process diagrams or flowcharts. A workflow diagram is a graphic depiction of the steps, sequence of steps, and flow control that constitute a process using standard symbols and conventions.

Index

Note: Page numbers followed by f and t refer to figures and tables, respectively.

E

T